The Most Important Clock in America

The David Rittenhouse Astronomical Musical Clock at Drexel University

The Most Important Clock in America

The David Rittenhouse Astronomical Musical Clock at Drexel University

Ronald R. Hoppes

American Philosophical Society
Philadelphia • 2009

Transactions of the
American Philosophical Society
Held at Philadelphia
For Promoting Useful Knowledge
Volume 99, Part 2

ISBN-13: 978-1-60618-992-4

US ISSN: 0065-9746

Photos and movement gear counts by conservator Bruce Forman
Other contributors: Eric Chandlee Wilson, Conservator
Jacqueline M. DeGroff, Curator, The Drexel Collection, Drexel University

---

Library of Congress Cataloging-in-Publication Data

Hoppes, Ronald R, 1935-
   The most important clock in America : the David Rittenhouse astronomical musical clock
at Drexel University / Ronald R. Hoppes.
      p. cm. — (Transactions series, ISSN 0065-9746 ; v. 99, pt. 2)
   Includes bibliographical references and index.
   ISBN 978-1-60618-992-4
   1. Clocks and watches—Pennsylvania—Philadelphia—Design and construction.
2. Astronomical clocks—Pennsylvania—Philadelphia—Design and construction.
3. Rittenhouse, David, 1732-1796. I. Title.
   TS543.5.P5H67 2009
   681.1'130974811—dc22
                              2009005296

Design and Typesetting by
International Graphics Services.

Printed in the United States of America by Hamilton Printing.

Contents

Foreword

When I assumed the presidency of Drexel University in 1995, my first task was to stabilize the University's precarious financial situation. Among the cash-raising measures suggested to me was the sale of our 18th-century David Rittenhouse astronomical musical clock, one of the finest examples of the genius of Philadelphia's preeminent astronomer, mathematician, and clockmaker.

The several million dollars such a sale would generate was a significant sum, given the condition of Drexel's finances, and in fact a sales catalog had already been prepared by a New York auction house. But then I looked at the clock.

Completed in 1773 by Rittenhouse (1732–1796), our clock represents the pinnacle of American clock making of the period. It provides the time, date, phase of the Moon, placement of the then-known planets, the equation of time, the position of the Sun and the Moon in the zodiac and the revolution of the Moon around the Earth. It also plays ten tunes. The clock came to Drexel in 1894 when the widow of George W. Childs, the best friend of our founder, Anthony J. Drexel, donated this masterpiece to The Drexel Collection.

As an engineer, I was filled with admiration for the genius and technical contributions of David Rittenhouse. As president, I knew how important the preservation of the historical and cultural integrity of Drexel would be to reversing the University's fortunes. And as an appreciative devotee of art and science, I became determined to keep this treasure on campus.

The clock stayed, and it has become a symbol of the tradition of Drexel University and the legacy of Anthony J. Drexel. In 2005 we were able to restore and conserve the clock thanks to a gift from Miriam and Brian O'Neill. I am pleased that the American Philosophical Society, founded by Benjamin Franklin and once presided over by David Rittenhouse himself, has published this book by clock conservator Ron Hoppes, a 1960 Drexel graduate.

I invite the reader to visit our Anthony J. Drexel Picture Gallery, where our remarkable David Rittenhouse astronomical musical clock is on display.

Constantine Papadakis, PhD
President, Drexel University

About the Author

Ronald R. Hoppes is a retired development engineer with a degree in electrical engineering from Drexel University. He holds patents in both the United States and Canada. He has been a member of the National Association of Watch and Clock Collectors since 1975 and has an interest in unusual escapements and mechanisms. He collects clocks, tools, and related books on theory and repair. Working in both wood and metal, he makes replacement parts for cabinets and movements that are faithful to the originals. He also presents workshops at NAWCC chapter 1 meetings.

Portrait of David Rittenhouse (1791), by Charles Willson Peale (1741–1827), Fine Arts Collection, American Philosophical Society.

Biography of David Rittenhouse

David Rittenhouse (1732–1796), the maker of Drexel University's Astronomical Musical Clock, was one of the most important scientists of the eighteenth century. He was a mathematician, astronomer, master craftsman of scientific instruments, land surveyor, and maker of astronomical clocks and orreries, which are mechanical models of the solar system. [1]

One of ten children, Rittenhouse was born on April 8th to farmers, Matthias and Elizabeth Rittenhouse, who lived along the Wissahickon Creek near Philadelphia. The family farmed the land where David's ancestor, Wilhelm Rittinghausen, built the first paper mill in the Colonies in 1690. When David was two years old his family moved to a farm in Norriton, Montgomery County, Pennsylvania.

Rittenhouse showed mathematical and mechanical abilities at an early age and acquired proficiency in observational, practical, and theoretical astronomy. David's siblings frequently found mathematical notations written on the fence posts in the fields surrounding the farm. Upon the death of his maternal uncle, David Williams, a carpenter who frequently visited the family, David inherited a box of tools and books that David referred to as a "treasure."[2]

By 1751, at the age of nineteen, Rittenhouse was established as a clockmaker and would sell his scientific instruments and clocks by the roadside. He trained his younger brother, Benjamin (1740–1825), in this art and the two young men became known as reputable clockmakers.

An important educational influence emerged in Rittenhouse's life in 1751 when Reverend Thomas Barton (1730–1780), an Episcopal clergyman and graduate of Trinity College, Dublin, settled in Norriton. Two years later, Barton married David's sister, Esther. Barton recognized the genius of Rittenhouse and provided him with scientific books and publications from the Colonies and

---

[1]The factual information was gathered from Brooke Hindle, *David Rittenhouse* (Princeton, NJ, Princeton University Press, 1964) and William Barton, *The Memoirs of David Rittenhouse, LLD.F.R.S.* (Philadelphia, Edward Parker, 1813).
[2]Hindle, 15.

Europe. Rittenhouse mastered an English translation of Sir Isaac Newton's
Principia, which became an inspiration for his orreries.

In 1766, having established his clock business and the owner of the family
farm, David married Eleanor Colston, called Nelly, a Quaker and daughter
of a neighboring farmer. They had two children: Elizabeth (Betsy) born in
1767, and Esther (Hetty) born in 1768.

Rittenhouse's brother-in-law, Thomas Barton, introduced him to Reverend
William Smith (1727–1803), the Provost of the College of Philadelphia (now
the University of Pennsylvania). Smith, active in the American Philosophical
Society, invited Rittenhouse to meetings at the Society and in 1768 Rittenhouse
was elected a member.

In anticipation of the transit of Venus in 1769, an important event that
would allow astronomers to calculate the distance between the Earth and
the Sun, David Rittenhouse created astronomical instruments and built an
observatory in Norriton. There would not be another opportunity to view the
transit of Venus for 105 years. The American Philosophical Society sponsored
the 1769 observation. Three locations were arranged: Rittenhouse's farm in
Norriton, State House Square in Philadelphia, and the Capes in Delaware.

Participating with Rittenhouse in Norriton were Provost William Smith,
John Lukens, and John Sellers. In Philadelphia at the public observatory in
State House Square were John Ewing, Thomas Shippen, Hugh Williamson,
Thomas Prior, Charles Thomson, and James Pearson. At the Lighthouse near
the Capes of Delaware, Owen Biddle and Joel Bailey observed this event.

Rittenhouse had predicted the initial contact of Venus with the Sun to be
at 2:11 p.m. Sixty seconds prior, Smith called out a warning for all to be
attentive. At this important moment, Rittenhouse fainted and was unconscious
for about six minutes. Dr. Benjamin Rush later stated of the event, "It excited
in the instant of one of the contacts of the planet with the Sun, an emotion
of delight so exquisite and powerful as to induce fainting."[3] Rittenhouse's
response to the second internal contact was described with full detail. However,
there were discrepancies from the observers in the three locations. Provost
Smith wrote to Thomas Penn in London praising the mathematical and
astronomical skills of David Rittenhouse. Irritated by Smith's competitive
approach, John Ewing wrote to Benjamin Franklin in London praising the
work of Owen Biddle and Joel Bailey.[4] Although an important step in their
astronomical studies was completed, there was never a consistent consensus
among the observers about the transit of Venus in 1769.

---

[3]Benjamin Rush, *An Eulogium Intended to Perpetuate the Memory of David Rittenhouse,*
Philadelphia, 1796, 12.

It should be noted that David Rittenhouse was in poor health all of his life. It is believed
that he suffered from a duodenal ulcer, which may have contributed to his fainting.

[4]Hindle, 58–63.

Several years prior, in 1767, Rittenhouse had begun the construction of an orrery, a mechanical model of the solar system. The term "orrery," was derived from a famous orrery built in England in 1713 for Charles Boyle, fourth Earl of Orrery. During the eighteenth century orreries were considered essential in teaching natural philosophy. Harvard procured one in 1732 and Lewis Evans used an orrery at the College of New Jersey (now Princeton University) in 1751. Rittenhouse's orrery was intended for the College of Philadelphia and details had been discussed with Provost Smith. In 1770, Reverend John Witherspoon, the newly appointed President of the College of New Jersey, along with a few Trustees, visited Rittenhouse. They praised him, admired the orrery, and asked Rittenhouse if they could buy it for the College of New Jersey. Rittenhouse agreed. There had been a rivalry between the two colleges and Provost Smith at the College of Philadelphia was irritated because Rittenhouse had not kept his promise. Rittenhouse wrote to Witherspoon to explain that the orrery was initially intended for the College of Philadelphia. Witherspoon never responded. Rittenhouse promised another orrery to the College of Philadelphia and delivered both orreries in 1771 to diffuse the resentment. Thomas Jefferson, a friend and admirer of Rittenhouse, said of the orrery at the College of Philadelphia, "He has not indeed made a world, but he has by imitation approached nearer its maker than any man who has lived from creation to this day."[5]

Provost Smith had been urging Rittenhouse to move to Philadelphia so he could participate more directly in the city's scientific community. In December 1770, Rittenhouse, his wife, and two young daughters moved to a house on the southwest corner of Seventh and Arch Streets in Philadelphia. Sadly, his wife, Eleanor, died in childbirth in February 1771. Two years later he married Hannah Jacobs. They had one daughter who died in infancy.

It was during this time that Rittenhouse built Drexel University's Astronomical Musical Clock, dating to c. 1773. This Rittenhouse clock has long been considered the most important clock in America. William Barton, son of Thomas and Esther Barton and nephew of Rittenhouse, wrote a biography in 1813, *The Memoirs of the Life of David Rittenhouse, LLD.F.R.S.*, and indicated that the clock was built for Joseph Potts of Philadelphia, who paid $640.00 for it. In the spring of 1774 the clock was purchased by Thomas Prior, an associate of David Rittenhouse, an active member of the American Philosophical Society, and participant in Philadelphia during the observation of the transit of Venus in 1769. Upon his death in 1801, the clock was next owned by Benjamin Smith Barton, nephew of Rittenhouse and brother of the biographer.[6] In 1815 the clock was acquired by James Swain of the *Philadelphia*

---

[5]Thomas Jefferson, *Notes on the State of Virginia*, 1782, 120.
[6]Barton, 203, n. 87.

Public Ledger. George W. Childs, best friend of Anthony J. Drexel, purchased the clock from Swain's estate in 1879. In 1894, the widow of George W. Childs donated the clock to Drexel Institute of Art, Science and Industry (now Drexel University).

The Drexel University Clock is distinctive because of the scientific information it provides: The orrery demonstrates the location of the six planets known at the time: Venus, Earth, Mars, Jupiter, Saturn, and Mercury; the lunarium indicates the phase of the Moon; a filigreed hand shows the month; and an aperture gives the date. The upper left dial shows the placement of the Sun and the Moon in the Zodiac; the upper right dial indicates the equation of time —the difference between solar time and mean time; the lower left dial shows the Moon's orbit around the Earth; and the lower right dial is a tune indicator for the ten songs the clock can play.

David Rittenhouse was a dedicated scientist as well as a committed patriot. The American Revolution began in Pennsylvania, where a strong radical party was active. He was a member of the Committee of Safety, the Pennsylvania General Assembly, the Pennsylvania Constitutional Convention, and Board of War. In 1779, he was named State Treasurer by George Washington. He served as Professor of Astronomy, Vice-Provost, and Trustee at the University of Pennsylvania.

Rittenhouse maintained his activities with the American Philosophical Society serving as Librarian, Secretary, and Vice-President. Following the death of Benjamin Franklin in 1790, Rittenhouse became President of the American Philosophical Society. In 1792, Rittenhouse was appointed by George Washington as the first Director of the United State Mint. In 1795 he was selected as a Foreign Member of the Royal Society of London.

David Rittenhouse enjoyed music, especially that of the composer, Franz Joseph Haydn. On Wednesday evenings, after Benjamin Franklin's return to Philadelphia from Paris in 1785, he would visit Franklin's home and they would listen to the music of Haydn. Francis Hopkinson, a signer of the Declaration of Independence, would play the harpsichord and when Thomas Jefferson was in Philadelphia, he would play the violin.[7] Rittenhouse's close friendship with these founding fathers is moving and attests to the stature of his character.

Rittenhouse was of poor health all of his life due to a duodenal ulcer. However, on June 22, 1796, he became gravely ill. He suffered an attack of cholera and realized he would not recover. He died on June 26, 1796, at the

---

[7]Julian P. Boyd, Ed., *The Boyd Papers*, "The Papers of Thomas Jefferson," Vol. 9, 322; Vol. 10, 249–250.

I am grateful to William Mills of Houston, Texas, for sharing his research on David Rittenhouse. Of particular importance is the information on the Wednesday evening gatherings at the home of Benjamin Franklin with David Rittenhouse, Francis Hopkinson, and Thomas Jefferson.

age of 64. His funeral was subdued, a few friends gathered at his home. A memorial service took place at the First Presbyterian Church on High Street in Philadelphia on December 17, 1796. President and Mrs. George Washington attended.[8]

David Rittenhouse's position as President of the American Philosophical Society was filled by Thomas Jefferson. While assuming the presidency Thomas Jefferson said the following, "Permit me to avail myself of this opportunity of expressing the sincere grief I feel for the loss of our beloved Rittenhouse. Genius, science, modesty, purity of morals, [and] simplicity of manners marked him as one of nature's best samples of the perfection she can cover under the human form."[9]

> Jacqueline M. DeGroff, Curator
> The Drexel Collection
> Drexel University

---

[8]Hindle, 364–365.
[9]Thomas Jefferson, "Early Proceedings of the American Philosophical Society," February 24, 1797, 251.

FIGURE 1.   The David Rittenhouse Astronomical Musical Clock
in Drexel University's Anthony J. Drexel Picture Gallery

1

Introduction

FIGURE 1 DEPICTS the clock as it appears today in the Anthony J. Drexel Picture Gallery. I first encountered this clock in the 1950s at Drexel University. It was on the second floor of the Main Building in the hall leading away from the atrium, which was referred to as the Main Court. I traveled the hall frequently between classes, some days almost embracing the clock while slowly advancing against an oncoming crowd of other students. During this slow advance I often looked at and pondered the many indications on the dial. In those days I knew nothing concerning the history of the clock, nor was I familiar with David Rittenhouse, whose name is on the dial. Surely a clock such as this must have many fascinating secrets to disclose. Perhaps some day, when I had more leisure time, I could find this information in some book and satisfy my curiosity.

Over the intervening years I have found little in the way of detailed descriptions of the dial and its indications, or the movement of the clock. Then one day many years later I received the opportunity to examine the clock during its cleaning and restoration. Now the information I had been seeking was before me and my curiosity was rekindled once again. With access to the interior of the clock, I now had the opportunity to discover and record in detail the movement and various dial indications. The purpose of this publication is to describe these indications and document the mechanical mechanism of this important historical clock.

Some time before 1767, Rittenhouse began designing an orrery. It was at this time that he built his first tall-case clock with an orrery in the arch above the dial. This first clock, built in 1767, is currently owned by the Pennsylvania Hospital. However, the strike and all of the gearing for the dial indications and orrery no longer represent Rittenhouse's work. Over the years they have been subjected to modification and alteration. At one point the gearing for the dial indications were lost. Replacement gearing has been made, however, it is considerably different from that of the original. Photographs of this clock, its dial, and movement can be found in *Clockmakers of Montgomery County* by Bruce Forman.

During the years of 1767–1770 Rittenhouse designed and produced two orreries, one for Princeton University and one for the University of Pennsylvania. Then in 1773 he built a second more elaborate clock with an orrery. This clock, currently owned by Drexel University, is a very good example of Rittenhouse's work, as it has experienced only minor changes over the years. This was the masterpiece of his clock-making career and is a national historical treasure. It has indications and mechanical features I have not seen on other clocks and the accuracy of the astronomical indications are excellent. I feel many of the features found in these two clocks are the result of his work on the orreries.

For instance, although the orrery and other indications follow the astronomical rates, the calendar indication of the clocks follows the Julian calendar, not the Gregorian calendar, which was in use then as well as today. The Julian calendar was chosen for the calendar indications because of the regularity of its leap-year rule to which there are no exceptions. With the 100-year rule of the Georgian calendar eliminated, the mechanical implementation of the calendar indication could be easily accomplished. The calendar date indications for the orries are similar, observing the Julian leap-year rules as the orrery simulates the astronomical rates. However, the orreries have an auxiliary dial that indicates when the user needs to correct the indicated Julian date for the Gregorian calendar's 100-year leap-year rule. The clocks do not have this additional dial, and the responsibility is placed on the owner to keep track of and correct for the 100-year rule.

Another unique feature is the method in which the equation of time has been achieved. The common approach is to translate the equation-of-time variance onto a kidney-shaped disk and use it to drive the indication. However, being an astronomer and mathematician with an understanding of the astronomical behavior that causes this variation, Rittenhouse chose to build a mechanism that generates the indication by simulating the astronomical actions. I have analyzed the mechanism and the astronomical movements to arrive at an understanding of how he produced this mechanism. I resorted to a computer analysis, however, not the laborious calculations with pencil and paper that were required of him.

The age markings he used on the moon dial are different from those of other clocks. Most clocks indicate the age of the Moon in days since the last new moon appeared. Rittenhouse chose not to indicate the age of the Moon, but the time at which each of the four phases occur as well as the number of days before or after the occurrence of a new and full moon. In those days it was customary to place a lot of prominence on the occurrences of the lunar phases and this provided a more accurate indication of when they occur.

I have partitioned the following description of the clock and movement into seven parts: the dial, the gear trains, an analysis of the calendar gear train, the equation-of-time indicator and its analysis, the strike operation, the operation of music train, and a procedure for setting the dial indications.

2

The Dial

A CLOSE-UP OF THE DIAL IS SHOWN in Figure 2. Descriptions of the various dial indications, their functions, and the dial selections follow.

The Orrery

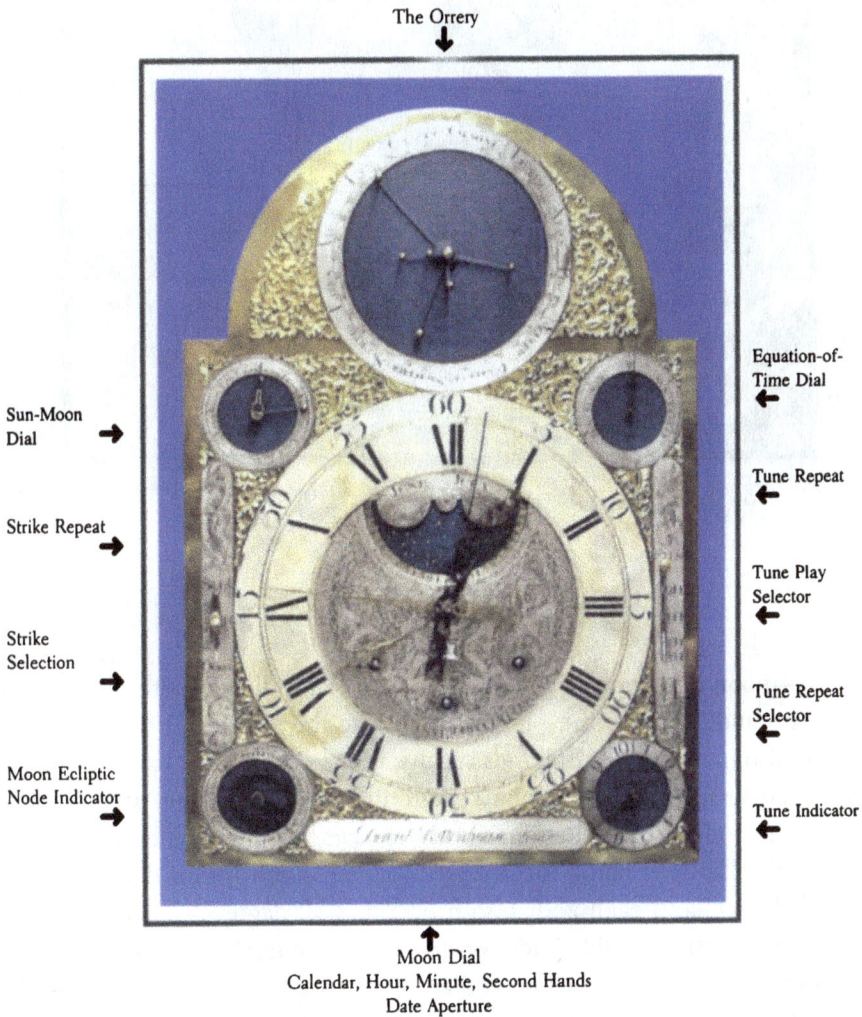

Sun-Moon Dial

Strike Repeat

Strike Selection

Moon Ecliptic Node Indicator

Equation-of-Time Dial

Tune Repeat

Tune Play Selector

Tune Repeat Selector

Tune Indicator

Moon Dial
Calendar, Hour, Minute, Second Hands
Date Aperture

FIGURE 2.   The Dial of the Clock

FIGURE 3.   Sun and Moon in Zodiac Dial

THE SUN–MOON DIAL

The sun–moon dial appears in Figure 3. This dial indicates both the position of the Sun and Moon as they appear relative to the zodiac. The phase of the Moon is also indicated. Both hands revolve in the counterclockwise direction.

The zodiac surrounds our solar system and the edge of this dial indicates the various zodiac constellations in the heavens. When reading the dial, consider the Earth as being located at its center. Each minor division of the dial is 1 degree, and each zodiac sign occupies 30 degrees.

The hand with the gold sphere on its tip represents the Sun and indicates the Sun's current position in the zodiac. Of course we cannot see the zodiac around the Sun because of the Sun's brilliance. The portion of the zodiac that we see at night is opposite the Sun. The hand indicating the Sun's position makes one revolution around the dial in one year. The other hand indicates both the Moon's position in the zodiac, and the phase of the Moon. The moon makes one revolution around the dial per lunar cycle, which is 27.322 days sidereal or 29.53 days synodical.

The sidereal rate is the time required to complete one revolution around the dial. The synodical rate is the observed appearance of one revolution of the Moon as seen from the Earth. Consider that the Sun travels around the zodiac in 365.25636 days, or 0.985609 degrees per day (360 deg per yr/ 365.25636 days per yr = 0.985609 deg per day). So during the 27.322 days of the sidereal revolution, the Sun has advanced an additional 26.9288 degrees (27.322 days × 0.985609 degrees per day = 26.9288 degrees). This advance requires an additional interval of 2.208 days (29.530 days −27.322 days = 2.208 days) for the Moon to arrive at the same relative position in the sky above the Earth.

The hand carries a representation of the Moon. One half of it is black to represent a new moon, whereas the other half is white to represent a full moon. As the hand progresses around the dial, it rotates to indicate the Moon's various phases. *Note: The red arrow points to a small square shaft at the lower edge of this dial. This is for moving the orrery by means of a crank. When it is placed on the shaft, a pin beside the shaft is depressed by a skirt on the crank. Depressing the pin disengages the orrery gears from the clock movement, allowing it to be operated manually by the crank. The clock's calendar hand moves as the orrery is cranked, indicating the date corresponding to the positions of the planets. This allows both past and future planet relationships to be examined and analyzed. The positions of both the sun and moon dial and the moon elliptic orbit and node indications also move accordingly.*

REPEAT AND STRIKE SELECTOR

The repeat lever can be seen in the upper left portion of Figure 4. It is a slider, which when moved upward and then returned to the normal down position repeats the last strike. The strike repeat will not function if the clock is preparing for the next strike, which begins approximately 5 minutes prior to a strike.

The strike selector is located directly below the trip lever, and is in the center left of Figure 4. Three selections are provided.

Silent — No hour or quarters are struck.

1H — Striking occurs on the hour.

Quary — Striking occurs on the quarter hour.

FIGURE 4. Repeat and Strike Selector

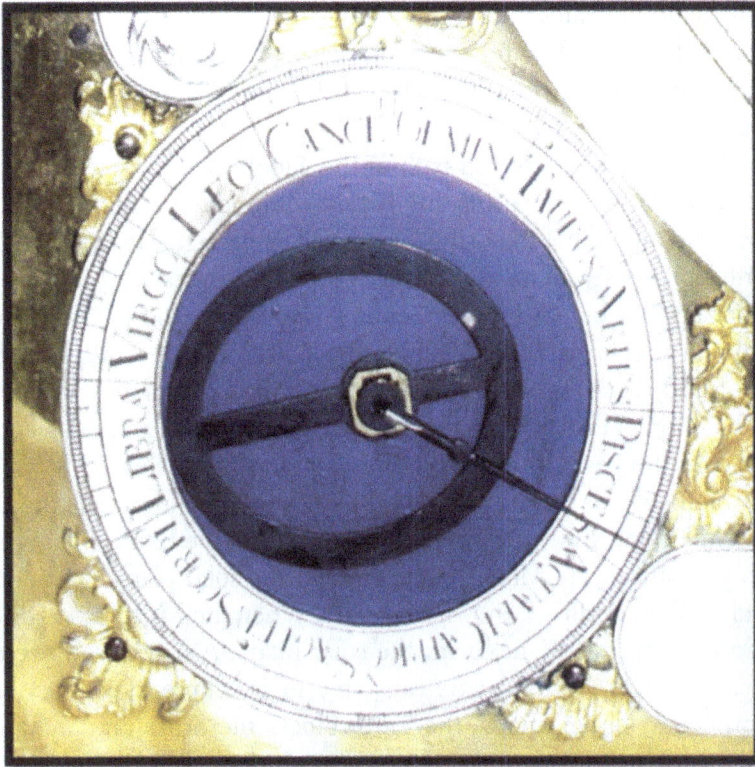

FIGURE 5. Moon Elliptic Orbit and Node Indicator

MOON ELLIPTIC ORBIT
AND NODE INDICATOR

The moon orbit indicator is shown in Figure 5. The edge of the dial represents the zodiac, similar to the sun–moon dial. The indication of this dial consists of an ellipse and a single hand. Consider the Earth to be located at the center of the dial, with the Moon orbiting counterclockwise around the Earth. The ellipse represents the orbit of the Moon around the Earth. Using the path of Earth's solar orbit as a reference plane, the Moon's orbit is inclined to it at an angle of 5.145 degrees. A more proper representation would be to place the ellipse at an angle of 5.145 degrees rather than parallel to the surface of the dial.

This inclination causes the orbital path to intersect the solar plane with one portion extending above the solar plain and another below it. The portion extending above the solar plane is known as the northern, or ascending, node. The portion below the solar plane is known as the southern, or descending, node. The gravitational attraction between the Sun and the Moon causes the nodes to shift westward (CW) at a rate of one revolution in 18.539688 years. The single hand indicates the mean longitude for the location of the northern high point of the ascending node. In addition to the shifting of the nodes, the elliptic path of the orbit also shifts. The elliptical orbit shifts eastward (CCW) one revolution in 8.787685 years.

From the Earth, the observed path of the Moon in the sky moves up and down in orbital latitude. By using this dial, in conjunction with the sun–moon dial above, the latitude of the rising Moon and its passage can be ascertained. The Earth's tilt causes an annual change of 23.45 degrees, plus the algebraic addition of up to + or − 5.145 degrees as a result of the tilt of Moon's orbit. So the annual angular variation is between 28.600 (23.450 + 5.150) and 18.300 (23.450 − 5.150). Although the 23.450 portion of the variation occurs annually, the 5.150 portion of the variation occurs over approximately an 18.6-year cycle, depending on the position of the elliptic path and the nodes.

A new moon always occurs when the Moon is on the Sun side of the Earth, whereas a full moon always occurs on the dark side of the Earth. The locations at which the path of the Moon passes through the solar plane from north to south and south to north are known as apsides. When the Moon passes through an apside it is directly aligned with the solar plane. A new moon occurring on an apside blocks the Sun's rays causing a solar eclipse to occur. When a full moon occurs on an apside, the Earth blocks the Sun's rays and a lunar eclipse occurs. The location of the apsides are not shown, but are approximately 90 degrees to the mean longitude hand.

EQUATION-OF-TIME INDICATOR

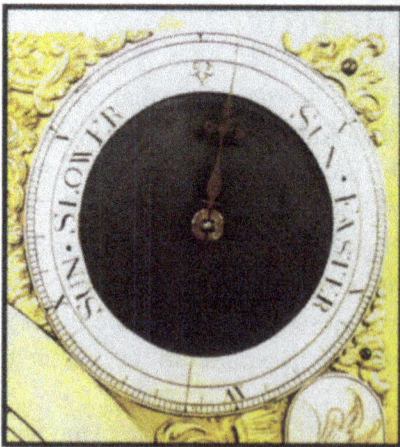

FIGURE 6. Equation-of-Time Dial

Figure 6 illustrates the equation-of-time indicator. It displays the difference between "true solar time" (sundial time) and "mean solar time" (mean solar time being clock time). In 1770 this was an important and very necessary piece of information for accurately setting the clock to the correct time. The reference for the determination of accurate time in that period was the garden sundial. If the sundial indicated 3:00 PM and the equation-of-time dial indicated the Sun to be four minutes slow on this calendar day, the correct clock timewas 3:04 PM and the clock hands were then set to indicate 3:04 PM.

In 1770, all places along the same longitude had the same mean solar time. However, for every different longitude there was a different mean solar time, resulting in different local times for each community. With the introduction of the railroads mean solar time was replaced with time zones varying by one-hour intervals. This greatly reduced the different local times a traveler would encounter on a journey. So to obtain true solar time with today's time zones, an additional factor must be added to the equation-of-time value. This value accounts for the difference between clock time, which is now time zone time, and what would be local mean solar time for the longitudinal location of the clock.

MANUAL TUNE PLAY

The knob in the upper right of Figure 7 is the trip lever for manual play of a tune. This is a slider, which, when moved downward and then returned to the up position, starts a repeat play of the selected tune.

FIGURE 7. Manual Play, Auto Play Interval, and Tune Repeat Selectors

TUNE-PLAY INTERVAL SELECTOR

This selector appears in the center right of Figure 7. There are four selections of how frequently tunes will be played:

Silent—No tunes will be played.
   2H—Music is played every 2 hours.
   1H—Music is played every hour.
  $^1/_2$H—Music is played every half hour.
  $^1/_4$H—Music is played every quarter hour.

TUNE-REPEAT SELECTOR

The tune-repeat selector appears in the lower right of Figure 7. It selects the number of times the tune is performed on each play:

1—Plays the tune once when play is initiated.
2—Plays the tune twice when play is initiated.
3—Plays the tune three times when play is initiated.
4—Plays the tune four times when play is initiated.

TUNE INDICATOR

The tune indicator appears in the lower right portion of Figure 8. It indicates which of the 10 tunes is presently selected for play.

The movement was designed to advance the tune selector one tune every 4 hours. The musical portion of this clock consists of two octaves of 15 bells.

FIGURE 8.   Tune Indicator

THE ORRERY

The orrery is located in the arch at the top of the dial, and indicates the position of the planets. Figure 9 illustrates the orrery. The zodiac around the dial denotes the positions of the planets in the sky. The rates of the planetary movements are:

FIGURE 9.   The Orrey

Center of dial—Sun
First planet from the Sun—Mercury
    87.98147 days/orbit CCW
Second planet from the Sun—Venus
    224.77318days/orbit CCW
Third planet from the Sun—Earth
    365.2564102days/orbit CCW
Fourth planet from the Sun— Mars
    687.0299 days/orbit CCW
Fifth planet from the Sun—Jupiter
    4,334.3761 days/orbit CCW
Sixth planet from the Sun—Saturn
    10,775.0641 days/orbit CCW

FIGURE 10. Date Aperture

THE CENTER OF THE DIAL

Date Aperture

The day-of-the-month aperture is illustrated in Figure 10. The day of the month appears in the aperture and is advanced one digit (1 thru 31) each day. The clock does not distinguish among months shorter than 31 days. The shorter months require a manual advance of the date at the end of the month. This is done by opening the door of the clock's waist, and reaching up behind the dial to manually advance the date wheel.

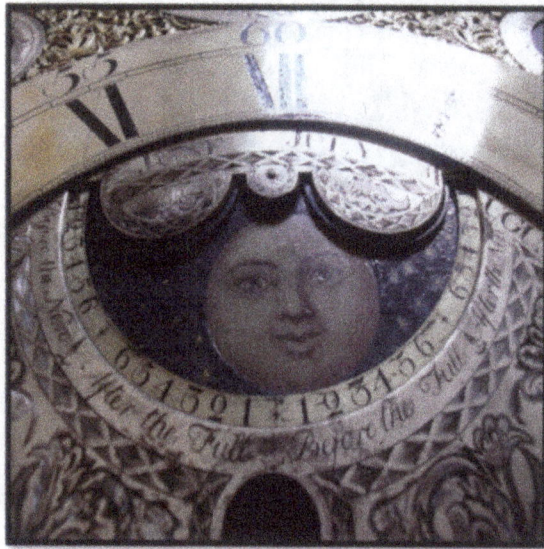

FIGURE 11. Moon Dial

THE MOON DIAL

The moon dial indicates the phase of the Moon and is located above the hand arbor, as illustrated in Figure 11.

Rather than indicating the age of the Moon as most other dials do, this dial indicates the days before and after the full and new moons. I think Rittenhouse chose this method because the occurrence of a new and full moon could be indicated and read more accurately.

The rate of this moon dial is 29.530588 days versus 29.5 days for the common tall-case moon dial. The error rate is 1 moon in 9,147 yrs versus 77 yrs for the common 29.5-day moon dial. Or to state it slightly differently, a 1% positional error of 0.6 days will occur in 91 yrs versus 9.5 months for the more common 29.5-day dial.

The dial has a friction slip drive, which allows manual positioning for corrections in the indication.

FIGURE 12. The Calendar Dial

THE CALENDAR DIAL

Figure 12 illustrates the complete calendar ring as it appears on the clock dial. The dial is engraved with the names of the 12 calendar months, each of which is divided into increments corresponding to the number of days in the month. The calendar hand does not move in discrete increments, but steadily advances during the 24-hour day.

Figure 13 is a close-up of the daily markings for the month of January. The markings of each month begin with a group of four days, followed by groups of five days, and a final group that varies in size depending on the number of days remaining in the month. The diagram in Figure 14 illustrates this more clearly.

FIGURE 13. The Month of January

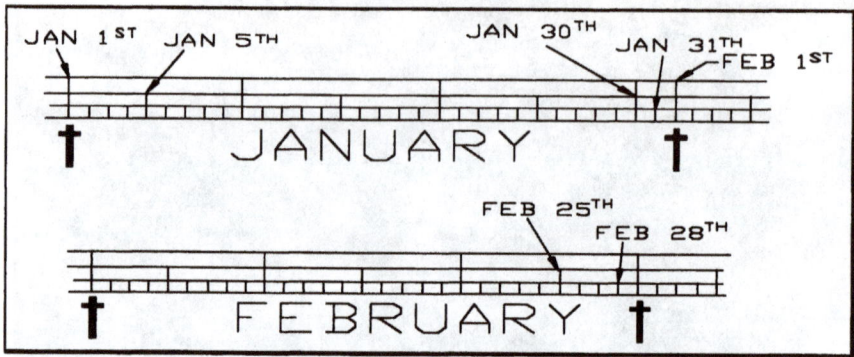

FIGURE 14. The Calendar Dial Markings

All months begin with day one (there is no day zero), the fourth division being the beginning of the fifth day. Using January as an example, it begins with a group of four days followed by five groups of five divisions each (five days each), for a total of thirty days. One must note that the day count, take day one, for example, continues for the entire day. Day two begins with the second division, and continues for the entire day until the pointer reaches the next division. What might seem to be an extra day at the end of January is actually the entire interval of day 31.

Months with thirty days end with a single division for the thirtieth day. Months with thirty-one days end with two divisions for the thirtieth and thirty-first days. February differs in that the fifth group consists of four days for a total of twenty-eight days.

It should be noted that the calendar hand revolves once in 365.25 days, accounting for the occurrence of leap years. Each year the daily indication is delayed by $1/4$ of a day, falling back into step on a leap year. Thus, the calendar hand follows the Julian calendar, whereas the Earth in the orrery follows the astronomical rates or Gregorian calendar. The difference between the two is the observance of the 100-year rule for leap year. The indications on his Princeton and University of Pennsylvania orrerys were similar. Because the Julian calendar indications are regular without any exceptions, it can be mechanized to indicate calendar dates without elaborate compilations. Because the 100-year rule occurs so infrequently, the complication of applying the 100-year adjustment (correction) to the calendar indication is placed on the user.

The leap-year rule for the Julian calendar is as follows: If the year is exactly divisible by four it shall be a leap year. The rule for the Gregorian calendar is: If the year is exactly divisible by four it shall be a leap year unless it is exactly divisible by 100, in which case it shall not be a leap year, unless it is exactly divisible by 400.

In certain astronomical work such as the observance of variable stars or when the time of occurrence of some phenomena is of much interest, Julian dates are frequently used because they are not complicated by the 100-year rule. The Julian date used is the count from January 1, 4713 BC, this date being the beginning of several subsidiary astronomical cycles.

3

The Gear Trains

THE RELATIVE POSITIONS, rotational rates, and tooth counts of the shafts and gears are given in the text that follows. Tooth counts on the movement have been obtained and verified by physical counts. Italicized items and tooth counts were calculated by the author, and though they are representative of the actual dial gearing, they have not been verified by physical counts. Rotational rates of dial indications are likewise based on the calculated counts, and although I believe them to be correct, they may not be in complete conformance with those of the clock. A photo of the gearing located between the plates, as well as a line drawing of the gearing is illustrated in Figure 15. The line drawing illustrates the designations and locations of each gear listed in Tables 1 through 4. Table 1 describes the time train gearing located along the central portion of the movement, whereas Table 2 describes the music train gearing located along the right-hand portion of the movement. Table 3 describes the strike train gearing located along the movement's left portion. An analysis of the number of hammer strikes required for the strike settings, Hourly and Quarterly, appear in Table 4. The analysis is performed twice. Once if the clock is assembled to strike the quarters as 0, 1, 2, 3 and as it strikes quarters today 1, 2, 3, 4. Because of modifications made over the years it is difficult to determine what the original intended sequence was. However, if the strike quarters are struck as 0, 1, 2, 3, the winding of both the strike and time trains can be performed together every 32 days. This fact is a strong indication that the clock originally did not strike four quarters after the hour strikes were completed.

FIGURE 15. Movement Gearing Between Plates

Table 1.   Time Train Gearing

Shaft	Tooth Counts	Diameters	Rate
			60 × 30 × 2 = 3600 beats/hr
T4-Escape	30/7	1.888/0.259	60 rev/hr
T3-3rd Shaft	84/7	2.634/0.256	5 rev/hr
T2-2nd Shaft	105/8	3.344/0.303	0.3333 rev/hr

(T2-2nd Shaft also drives motion work and gears on front of movement)

| T1-Winding Shaft | 128/2" Diam (16 groves) | 4.057/2 | 0.020833rev/hr |

Weight Drop: 2" × π × 0.020833rev/hr × 1/2 = 0.130898"/hr
24hrs × = 0.130898"/hr = 1.57077"/day
Maximum weight drop:2" × π × (16 groves) = 100.5301" × 1/2(for compound pulley) = 50.2655" drop.
50.2655"/(1.57077"/day) = **32 days max between windings**

Table 2.   Music Train Gearing

Shaft	Tooth Counts	Diameters	Rate
M4-Fly	8		147 rev/tune
M3-3rd Shaft	84/6	2.532/0.272	14 rev/tune
M5-Cylinder	84 teeth on cylinder		1 rev/tune (~ 0.25 sec)
M2-2nd Shaft	84/7	3.078/0.262	1 rev/tune
M1-Winding Shaft	160/2" Diam (19 groves)	4.914/2	0.04375 rev/tune

Weight Drop: 2" × π × 0.04375 rev/tune × 1/2 = 0.1374447"/tune
Maximum weight drop: 2" × π × (19 groves) = 119.38" × 1/2(for compound pulley) = 59.69" drop.
Or (59.69" drop/0.1374447"drop/tune) = 434.28 tunes at 12 per day (every 2 hrs.) = 36 days at 1 repeat.
24 per day (every hr.) = 18.1 days at 1 repeat.
48 per day (every 1/2 hr.) = 9.0 days at 1 repeat.
96 per day (every 1/4 hr.) = 4.5 days at 1 repeat.
1.1 days at 4 repeats.
The tune can be repeated 1, 2, 3, or 4 times. The tripping and playing of the tune and the tripping of the strike are separate and are not interlocked.

Table 3. Strike Train Gearing

Shaft	Tooth Counts	Diameters	Rate
S5-Fly	6	0.233	1152 rev/28 strikes
S4-4th Shaft	48/7		144 rev/28 strikes
S3-3rd Shaft	72/7	3.264/0.267	14 rev/28 strikes
S2-2nd Shaft	98/8 (28 pins for strike)	3.41/0.35	1 rev/28 strikes
S1-Winding Shaft	128/2" Diam (16 groves)	4.25/2	0.0625 rev/28 strikes

Weight Drop: $2'' \times \pi \times 0.0625$ rev/28 strikes $\times 1/2 = 0.1963495''$/28 strikes
0.1963495"/28 strikes or 0.00701248"drop/strike

Maximum weight drop: $2'' \times \pi \times (16$ groves$) = 100.5301'' \times 1/2$(for compound pulley)
= 50.2655" drop.

Or (50.2655" drop/0.00701248"drop/strike) = **7168 strikes maximum per winding.**

Table 4. Hammer Strikes per 12 hours

Hours only	12+11+10+9+8+7+6+5+4+3+2+1 = 78 Hour hammer blows in 12 hours.
1st, 2nd, & 3rd quarters	12 × (1 + 2 + 3) = 72 blows in 12 hours.
1st, 2nd, 3rd, & 4th quarters	12 × (1 + 2 + 3 + 4) = 120 blows in 12 hours.

For Hours Only
      7168 strikes/{78 × 2 every 24 hrs} = 45 days

For Hours and 1/4-, 1/2-, 3/4-hr strikes
      7168 strikes/{(78 +72) × 2 every 24 hrs} = 32.29 days

For Hours and 4/4 strike following the hour strike
      7168 strikes/{(78 +48) × 2 every 24 hrs} = 28.4 days

For Hours and 1/4-, 1/2-, 3/4-, & 4/4-hr strikes
      7168 strikes/{(78 +120) × 2 every 24 hrs} = 18.1 days

GEARING ON THE FRONT PLATE

Figure 16 is a photograph of the gearing on the front plate of the movement. Figure 17 is a line drawing of the same gearing, which also shows the flow of power transmission among the gear trains.

Motive power is supplied to the gearing on the front plate by the T2 shaft of the movement, which extends through the plate. The T2 shaft is shown as a solid dot in Figure 17. The connecting lines indicate the transmission of motive power along the gear trains.

The diagram is repeated in Figure 18 without the labels, enabling it to be presented in a larger format. Figure 18 also identifies the location and position of each gear shown in Table 5.

FIGURE 16.   Gearing on the Front Plate

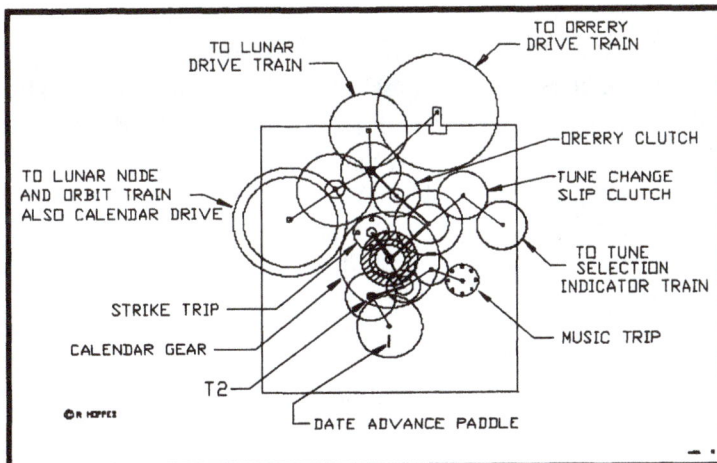

FIGURE 17.   Front Plate Gear Functions and Power Transmission Flow

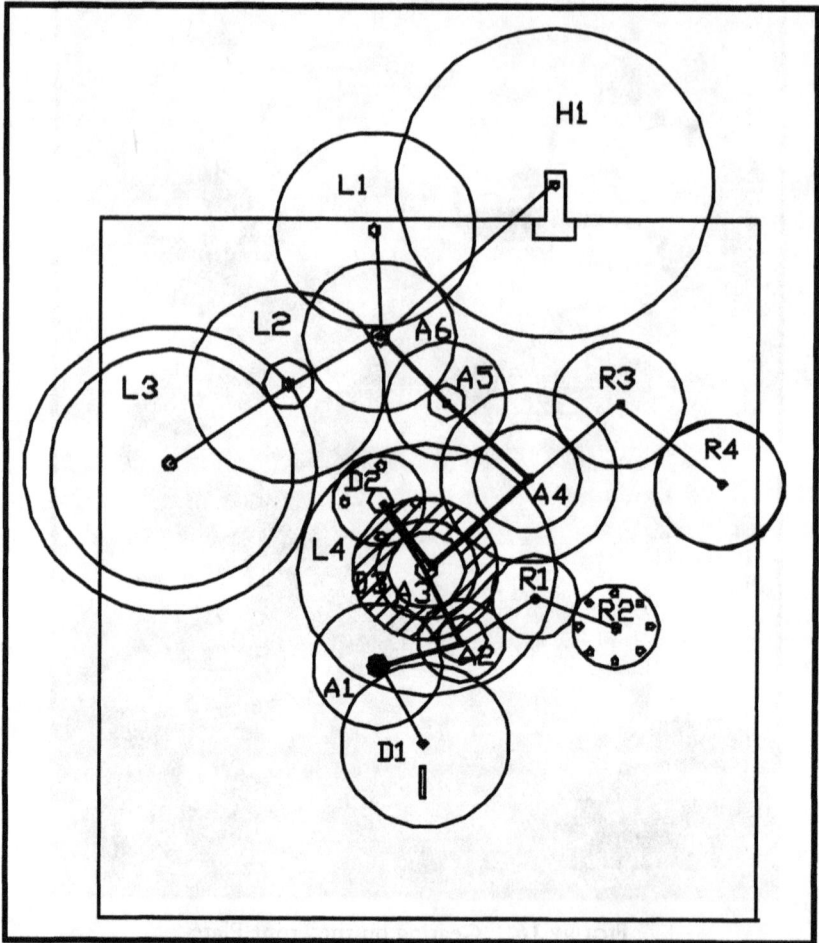

FIGURE 18.   Front Plate Gearing Power Flow

Gear A1 on the T2 shaft drives A2, which in turn drives A3. The minute, hour, and calendar elements are separate coaxial mounted wheels. The shaded circle on the center shaft, A3, is the minute wheel and quarter-hour snail. A3 drives D2, which drives D3. The crosshatched portion of D3 is the hour wheel and hour snail. Teeth on the back of D3 drive A4. The clear outer portion is the calendar wheel, L4, driven by L3. The L3 gear consists of two gears, one slightly smaller in diameter than the other. The power transmission path is A4, A5, A6, L2, L3, L4.

THE DIAL GEAR TRAINS

Figure 19 is a photograph of the gearing on the rear of the clock dial. All dial indications are driven by the time train of the clock movement. Although the

Table 5.   Front Plate Gearing

Shaft	Tooth Counts	Ratio	Diameters	Rate
Date Indicator				
A1-2nd Shaft	54T/8L	-	3.489	0.3333 rev/hr
D1-Date Paddle	64T	8/64	2.040	24 hrs/rev
Minute Hand				
A1-2nd Shaft	54T/8L -	3.489		0.3333 rev/hr
A2-Intermediate	36T/18L	54/18	1.365/0.694	1.00 rev/hr
A3-Min Shaft	36T	36/36	1.363	1.00 rev/hr
Strike Trip				
A3-Min Shaft	36T	36/36	1.363	1.00 rev/hr
D2-Strike Trip	36T/6L	36/36	1.364/0.268	1.00 rev/hr
Hour Hand				
D2-Strike Trip	36T/6L	36/36	1.364/0.268	1.00 rev/hr
D3-Hr Wheel/Snail	72T/30L	6/72		12 hr/rev
Music Trip				
A1-2nd Shaft	54T/8L	-	3.489	0.3333 rev/hr
A2-Intermediate	36T/18L	54/18	1.365/0.694	1.00 rev/hr
R1-Idle	36T	18/36	1.306	2 hr/rev
R2-Music Trip	36T	36/36	1.303	2 hr/rev
Auto Tune Change				
D3-Hr Wheel/Snail	72T /30L	6/72		12 hr/rev
A4-Intermediate	60T/36L	30/60	2.480/1.305	24 hr/rev
R3-Slip Clutch	51T/51T	36/51		34 hr/rev
R4-1st Tune Chg	60T	51/60		40 hr/rev

There are 10 tune selections/rev of R4 (6 teeth/tune) 4 hr/tune

Shaft	Tooth Counts	Ratio	Diameters	Rate
Lunar Drive Train				
A4-Intermediate	60T /36L	30/60		24 hr/rev
A5-Orrery Clutch	48T/16L	36/48	1.730/0.698	32 hr/rev
A6-Dial Functions	64T/12L	16/64	2.534/0.546	128 hrs/rev
L1-Dial Lunar Drive	62T	12/62	2.491	661.333 hrs/rev
Orrery Drive Train				
A4-Intermediate	60T /36L	30/60		24 hr/rev
A5-Orrery Clutch	48T/16L	36/48	1.730/0.698	32 hr/rev
A6-Dial Functions	64T/12L	16/64	2.534/0.546	128 hrs/rev
H1-Orrery Drive	78T	64/78	3.136	156 hrs/rev
Lunar Node Train				
A4-Intermediate	60T /36L	30/60		24 hr/rev
A5-Orrery Clutch	48T/16L	36/48	1.730/0.698	32 hr/rev
A6-Dial Functions	64T/12L	16/64	2.534/0.546	128 hrs/rev
L2-1st Lunar	72T/6L	12/72	2.843/0.321	768 hrs/rev
L3-2nd Lunar & Cal	73T/97L	6/73	3.739/2.911	9344 hrs/rev or 389.333 days/rev

(continued)

Table 5. Front Plate Gearing (*continued*)

Shaft	Tooth Counts	Ratio	Diameters	Rate
Calendar Hand				
A4-Intermediate	60T/36L	30/60		24 hr/rev
A5-Orrery Clutch	48T/16L	36/48	1.730/0.698	32 hr/rev
A6-Dial Functions	64T/12L	16/64	2.534/0.546	128 hrs/rev
L2-1st Lunar	72T/6L	12/72	2.843/0.321	768 hrs/rev
L3-2nd Lunar & Cal	73T/97L	6/73	3.739/2.911	9344 hrs/rev or 389.333 days/rev
L4-Cal Gear	91T	97/91	3.489	8766.020619 hrs/rev or 365.2508591 days/rev

The error rate is 365.256360416 days/yr − 365.2508591 days/yr = +0.0055013 day/yr 1/0.0055013 day error/yr =181.77 yrs/1 day error.

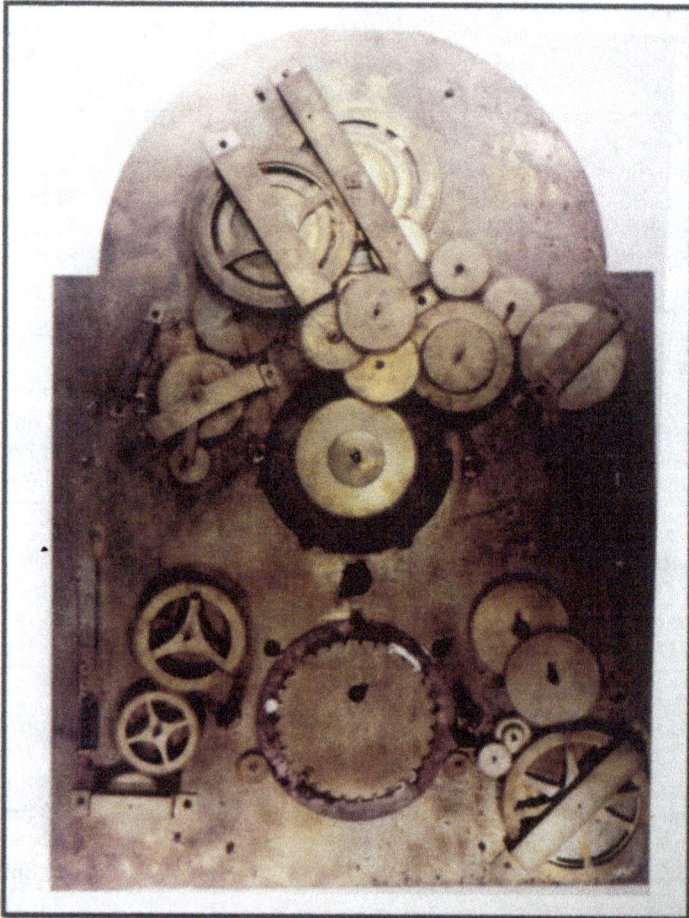

FIGURE 19. Gearing on Rear of Dial

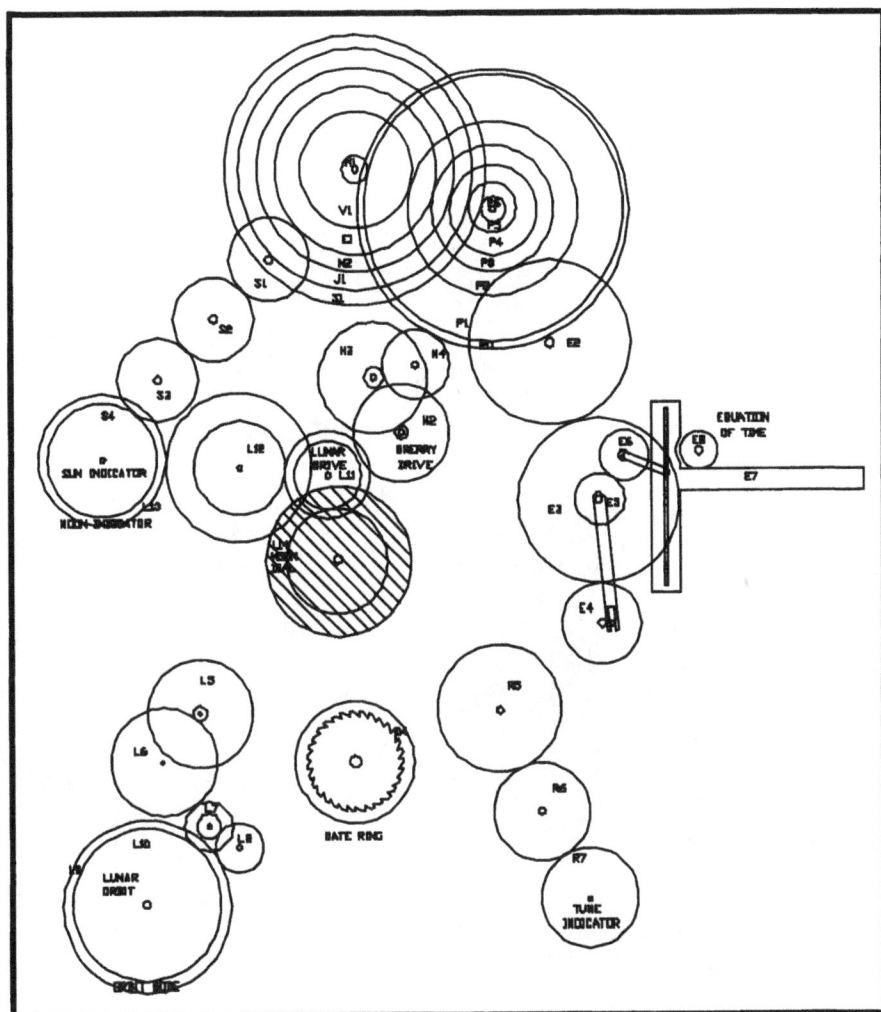

FIGURE 20. Diagram of Dial Gearing

dial has many gears and appears to be rather complex, it is no more complicated than many clock movements. A line diagram of the gearing is shown in Figure 20. Note that the diagram is drawn as if one were at the front of the clock looking at the gearing through the dial. This was done so that it could be combined with the gearing diagram of the movement front plate shown in Figure 21. Because of this view, Figure 20 is a mirror image of the photograph in Figure 19. Figure 20 also identifies the designation of each gear.

Figure 22 illustrates the flow of mobilizing power through the gearing of the dial. Note that there are several distinct separate power flow paths. There are the planetary paths of H2, H3, H4, P0 and P3, E1, S1, S2, S3, S4 and P3, E2; the equation-of-time gearing E3, E4, E6; as well as the orrery gearing

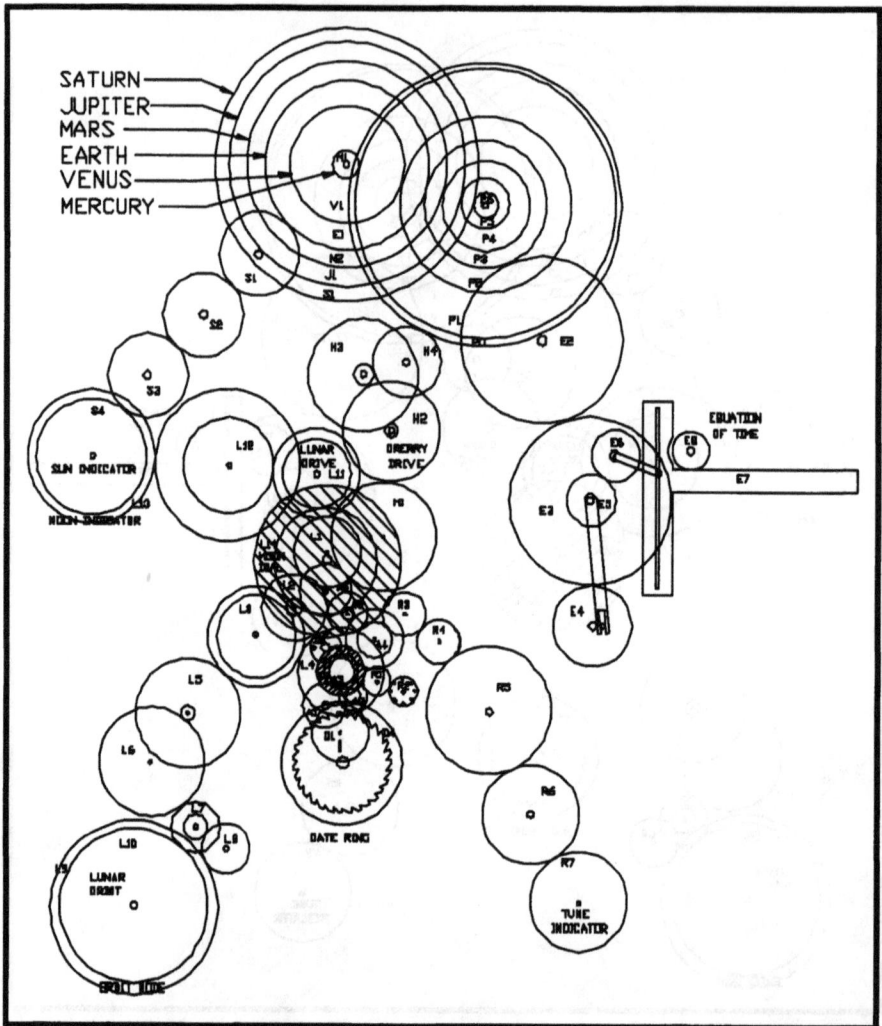

FIGURE 21. Composite Diagram of Movement Front Plate and Dial Gearing

P0 thru P6 and M1, V1, E1, M2, J1, S1. Then there are the lunar gearing paths of L11, L12, L13, and L5, L6, L7, L8, L9, L7, and L10. Lastly, the gearing R5, R6, R7 is for tune indication.

The equation-of-time gear, E5, is a free and separate gear on the same shaft with E3, to which a rigidly attached lever extends down to E4. E5 does not rotate but an oscillatory motion is imparted to it via the lever, caused by the rotary motion of E4. The shaft of E6 is mounted on E3 and driven around E5. The A pin in the end of the arm, which is rigidly attached to E6, moves E7 by its engagement in the slot. Both E7 and E8 are mounted close to the dial, placing them in another plane so as not to interfere with the other gears.

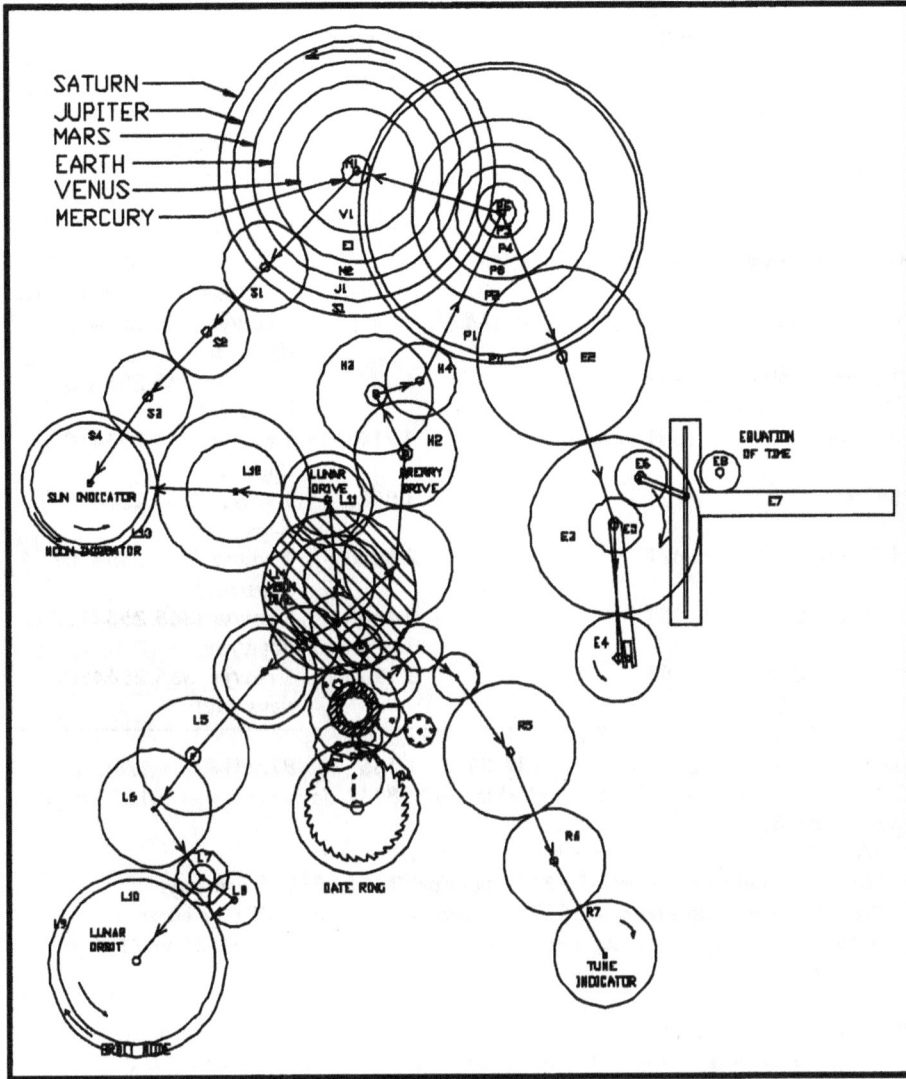

FIGURE 22. Diagram of Dial Power Transmission

The details of rotational rates and tooth counts are detailed in Table 6. Note that the planetary gears P0, P1, P2, P3, P4, P5, and P6 are all stacked and attached to each other in pyramid style, causing all of these gears to rotate as one assembly on the shaft. The assembly is driven by H4, which engages gear P0 of the stacked pyramid assembly. The gears M1, V1, E1, M2, J1, S1 are also mounted on a single shaft but are separate, each rotating at a rate determined by its ratio to its mating gear in the pyramid. A photograph of the M1, V1, E1, M2, J1, and S1 gears appears in Figure 23.

Table 6. Dial Gearing

Shaft	Tooth Counts	Ratio	Diameters	Rate
H1-Orrery Driv	78T	64/78	3.136	156 hrs/rev (6.5 days/rev)
H2-Orrery Drive	60T/12L	78/60		120 hrs/rev (8.45 days/rev)
H3-Orrery Drive	77T/13L	12/77		770 hrs/rev (32.08333333 days/rev)
H4-Orrery Drive	48TIdler	13/48		2843.0769 hrs/rev (118.4615385 days/rev)
P0-Orrery Unit	148T	48/148		1 rev/yr (365.2564102 days/rev)
P1-Orrery Unit	137T		4.835	1 rev/yr (365.2564102 days/rev)
P2-Orrery Unit	104T		3.714	1 rev/yr (365.2564102 days/rev)
P3-Orrery Unit	72T		3.000	1 rev/yr (365.2564102 days/rev)
P4-Orrery Unit	42T		2.083	1 rev/yr (365.2564102 days/rev)
P5-Orrery Unit	15T		0.466	1 rev/yr (365.2564102 days/rev)
P6-Orrery Unit	8T		0.196	1 rev/yr (365.2564102 days/rev)

P1/M1-Mercury   33T   137/33   1.165   87.98147 days/orbit
The error rate is 87.96934 days/orbit – 87.98147 days/orbit = –0.01213 days error/orbit.
1/0.01213 days error/orbit = 82.44 orbits/1day error.
(365.256360416 days/yr)/87.98147days/orbit = 4.151515 orbits/yr
(82.44 orbits/1day error)/4.151515 orbits/yr = 19.86 yrs/1 day error.
A 1% positional error is an error of 0.88 days occurring in 19.86 yrs/1 day error × 0.88 day/1% = 17.5 yrs/1%.

P2/V1-Venus   64T   104/64   2.286   224.77318 days/orbit
The error rate is 224.70069 days/orbit – 224.77318 days/orbit = –0.07249 days error/orbit.
1/0.07249 days error/orbit = 13.795 orbits/1day error.
(365.256360416 days/yr)/224.77318 days/orbit = 1.63 orbits/yr.
(13.795 orbits/1day error)/1.63 orbits/yr = 8.46 yrs/1 day error.
A 1% positional error is an error of 2.25 days occurring in 8.46 yrs/1 day error × 2.25 day/1% = 19.0 yrs/1%.

P3/E1-Earth   72   72/72   3.000   365.2564102 days/orbit
The error rate is 365.256360416 days/yr – 365.2564102 days/yr = –0.0000498 days error/yr.
1/0.0000498 days error/yr = 20,080.32 yrs/1day error.
A 1% positional error is an error of 3.65 days occurring in 20,080 yrs/1 day error × 3.65 day/1% = 73,292 yrs/1%.

Table 6. (continued)

P4/M2-Mars   79   42/79   3.917   687.0299 days/orbit
The error rate is 686.96 days/orbit – 687.0299 days/orbit =-0.0699 days error/ orbit.
1/0.0699 days error/orbit = 14.31 orbits/1day error.
(365.256360416 days/yr)/687.0299 days/orbit = 0.532 orbits/yr.
(14.31 orbits/1day error)/0.532 orbits/yr = 26.90 yrs/1 day error.
A 1% positional error is an error of 6.87 days occurs in 26.9 yrs/1day error × 6.87 days/1% = 184.8 yrs/1%.

P5/J1-Jupiter   178   15/178   5.534   4334.3761 days/orbit
The error rate is 4333.2867 days/orbit – 4334.3761 days/orbit = –1.089368 days/orbit.
1/1.089368 days error/orbit = 0.9180 orbits/1day error.
(365.256360416 days/yr)/4334.376 days/orbit = 0.8427 orbits/yr.
(0.9180 orbits/1day error)/0.8427 orbits/yr = 1.09 yrs/1 day error.
A 1% positional error is an error of 43.34 days occurs in 1.09 yrs/1day error × 43.34 days/1% = 47.24 yrs/1%.

P6/S1-Saturn   236   8/236   5.803   10775.0641 days/orbit
The error rate is 10756.1995 days/orbit – 10775.0641 days/orbit = – 18.86460 days/orbit.
1/18.86460 days error/orbit = 0.053 orbits/1day error.
(365.256360416 days/yr)/10775.0641 days/orbit = 0.033898 orbits/yr.
(0.053 orbits/1day error)/0.033898 orbits/yr = 1.56 yrs/1 day error.
A 1% positional error is an error of 107.8 days occurs in 1.56 yrs/1day error × 107.8 days/1% = 168.2 yrs/1%.

Sun in Zodiac

Shaft	Tooth Counts	Ratio	Diameters	Rate
E1-Earth	72T	72/72		1 rev/yr
S1-Sun Gear 1	52T	72/52		1.384615385 rev/yr
S2-Sun Gear 2	52T	52/52		1.384615385 rev/yr
S3-Sun Gear 3	52T	52/52		1.384615385 rev/yr
S4-Sun Indicator	72T	52/72		1 rev/yr

The error is the same as that for E1-Earth above.

Moon in Zodiac

Shaft	Tooth Counts	Ratio	Diameters	Rate
L1-Dial Lunar Drive	62T	12/62	2.491	661.333 hrs/rev
L11-Lunar Gear	69T/54L	62/69		735.999 hrs/rev
L12-Lunar Gear	62L/118T	69/62		735.999 hrs/rev
L13-Moon Indicator	117T	118/117		655.728 hrs/rev sidereal (27.32202 days/rev)

The error rate is 27.32202 days/sidereal rev – 27.32166 days/sidereal rev = 0.00036 day/sidereal rev.
1/0.00036 day/sidereal rev = 2777.7778 sidereal rev (moons)/–1 day error.
(364.25 sidereal days/yr)/27.32166 sidereal days/moon = 13.3319 sidereal moons/yr.
(2777.7778 sidereal moons/-1day error)/13.3319 sidereal moons/yr = 208.4 yrs/1 day error.

(continued)

Table 6. *(continued)*

Moon Dial

Shaft	Tooth Counts	Ratio	Diameters	Rate
L1-Dial Lunar Drive	62T	12/62	2.491	661.333 hrs/rev
L11-Lunar Gear	69T/54L	62/69		735.999629 hrs/rev
L14-Moon Dial	104T	54/104		1417.480767 hrs/rev

For a two-moon dial the rate is 59.061176 days/rev = 1417.468224 hrs/rev. The error rate for a two-moon dial is 1417.468224 hrs/rev – 1417.48076 hrs/rev = –0.01253 hrs/rev.
Or 1/2 rev × 0.01253 hrs/rev = 0.006268 hrs for one moon.
An indication error of one moon will occur in 1/0.006265 hrs/moon = 159.6 moons/1hr error.
At 12.3687466 moons/yr an error of one moon would occur in 9,145 years (159.6 moons/1hr error × 24 hrs/1 day × 29.530588 days/moon)/(12.3687466 moons/yr =12.9yrs). So a 1% error in the moons indicated position would occur in 91.45 years.
A 1% position error is 0.01 × 29.530588 days/moon = 0.29536 days, or an error of 0.29536 days × 24 hrs/day =7.09 hrs.

Lunar Orbit

Shaft	Tooth Counts	Ratio	Diameters	Rate
L3-2nd Lunar & Cal	73T/97L	6/73	3.739/2.911	9344 hrs/rev, 389.333 days/rev
L5-Orbit 1	92/13L	97/92		369.2642887 day/rev
L6 Orbit 2 (Idler)	81T	13/81		2300.800568 day/rev
L7-Orbit 3	32T/20L	81/32		908.958249 days/rev
L10-Lunar Orbit	134T	32/113		3209.758817 days/rev (8.787687685 yrs/rev)

The error is 8.8504 yrs/rev – 8.787685 yrs/rev = 0.062715 yrs/rev.

Orbit Node Longitude Hand

Shaft	Tooth Counts	Ratio	Diameters	Rate
L3-2nd Lunar & Cal	73T/97L	6/73	3.739/2.911	9344 hrs/rev or 389.333 days/rev
L5-Orbit 1	92T/13L	97/92		369.2642887 day/rev
L6-Orbit 2 (Idler)	81T	13/81		2300.800568 day/rev
L7-Orbit 3	32T/20L	81/32		908.958249 days/rev
L8-Orbit 4	32T	20/32		1454.333198 days/rev
L9-Orbit Node	154T	32/149		6771.738955 days/rev (18.539688 yrs/rev)

The error is 18.5996 yrs/rev – 18.539688 yrs/rev = 0.059912 yrs/rev.

Equation of Time

Shaft	Tooth Counts	Ratio	Diameters	Rate
P3-Orrery 3	72T	1/1		1 rev/yr
E2-E Drive 1	72T	1/1		1 rev/yr
E3-E Drive 2	72T	1/1		1 rev/yr
E4-Elliptic 2	36T	2/1		2 rev/yr
E5-Adder	30T	—		1 rev/yr
E6-Elliptic 1	30T	1/1		1 rev/yr
E7-S/F Drive	41T	—		47 sec/tooth
E8-Time S/F	40T	—		1880 sec/rev or 47 sec/tooth (Dial –14 min. 55 sec to +16 min 25 sec)

Table 6. *(continued)*

Tune Indicator			
R4-Tune Chg	**60T**	**51/60**	40 hr/rev (0.6 rev/day)
R5-2nd Tune Chg	**100T**	**60/100**	**24 hr/rev (1 rev/day)**
R6-3rd Tune Chg	**75T**	**100/75**	**18 hr/rev**
7-Tune Chg Ind	**60T**	**75/60**	**14.4 hr/rev (0.6 rev/ day)**

NOTES

1. Items in bold italics in the dial gearing tabulations were calculated by the author and have not been verified by physical counts. However, I believe that they are highly representative of the actual gearing. Rotational rates of dial indications are based on the calculated counts, and may not be in complete conformance with those of the clock. However, the error rates suggest that the calculated ratios are favorable representations of those in the clock.

2. Tooth counts on the movement have been obtained and verified by physical counts.

FIGURE 23.   The Orrery Gearing

4

The Calendar Hand

FIGURE 24 ILLUSTRATES the calendar hand gearing located on the front of the clock movement. The first gear of the train is driven by gear D3 on the hour shaft of Figure 18. I have searched libraries for information on calculating gear trains for astronomical applications. Of the few I have been able to locate, none of them approached the accuracy obtained by David Rittenhouse. Records and descriptions of his method, if they existed, are most likely lost. However, examination of the gear train used for the calendar hand reveals some interesting clues as to how he may have selected the ratios.

The gear train is driven by the hour shaft of the movement, which revolves twice per day. With this as the starting point, the tooth counts and rate of revolution of the gearing are shown in Table 7. Note that the rotational rate of the calendar hand is not 365 days, but closely matches the annual rate of the Earth around the sun. Thus the calendar gear train accounts for Julian leap years.

If the accepted Julian rate of the Earth around the Sun is 365.25 days, the gearing error is 0.0008591 days/yr, or 1/0.0008591 = 1164 years to accumulate an error of one day. However, we know the true rate to be 365.256360416 days, which results in an error of 365.2563604 − 365.2508591 = 0.0055013 days/rev or 161.8 years to accumulate an error of one day.

The accuracy of the rotational rates achieved by Rittenhouse's gearing indicates that he was well versed in mathematical calculations, far beyond other clockmakers of his time, and even most clockmakers working today.

Note that the first four ratios (30/60, 48/16, 64/12, 12/72) are convenient whole fractions, indicating they were likely derived by dividing 365.25 by a series of single-digit whole numbers. However, the ratios for the last two shafts were not produced by such a method, indicating they were arrived at by a different technique.

35

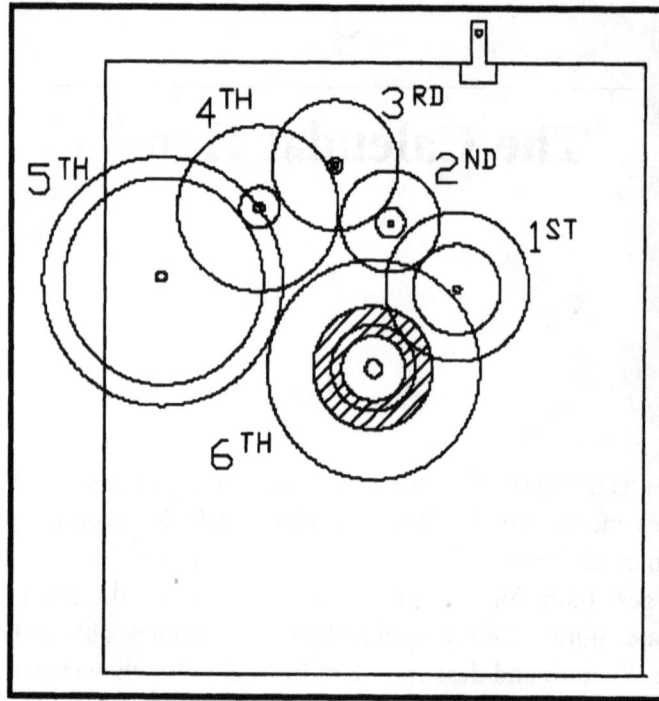

FIGURE 24. The Calendar Gear Train

Table 7. Calendar Gearing

Shaft	No. Teeth	Ratio	Decimal Ratio	Rate of Rotation	1/Rate of Rotation
Hour	30			2 rev/day	0.5 day/rev
1st	60/36	30/60(or 1/2)	0.5	1 rev/day	1 day/rev
2nd	48/16	36/48(or 3/4)	0.75	0.75 rev/day	1.33333333 day/rev
3rd	64/12	16/64(or 1/4)	0.25	0.1875 rev/day	5.33333333 day/rev
4th	72/6	12/72(or 1/6)	0.166666666	0.03125 rev/day	32 day/rev
5th	73/97	6/73	0.08219178	0.002568493151 rev/day	389.33333333 day/rev
6th	91	97/91	1.065934066	0.002737844347 rev/day	365.2508591 day/rev

After much thought, I believe the most likely approach used in determining the final gears was a combination of first calculating the required ratio to complete the gear train as a decimal value. This decimal value was then most likely used to select ratios that best fit the requirements from a ratio table.

Given the accuracy of his gearing I believe such a method was used for selecting the final ratios of this gear train.

Such a ratio table can be constructed with a series of columns and rows as shown in Table 8. Each column represents a tooth count for the denominator portion of the ratio, whereas each row represents the tooth count of the numerator. The rows and columns of the body of the table are then filled with the decimal equivalents of the ratio values.

One might argue that Rittenhouse may not have had such a table. However, with paper, pencil, and sufficient time, such a table could be generated. The entire table need not be constructed. Ratios in the table greater than 1 are

Table 8.   Ratios Expressed in Decimal Equivalents

Teeth Pinion/Gear	X/100	X/99	X/98	X/97	X/96	X/95
100/X	1	1.01010101	1.02040816	1.03092784	1.04166667	1.05263158
99/X	0.99	1	1.01020408	1.02061856	1.03125	1.04210526
98/X	0.98	0.98989899	1	1.01030928	1.02083333	1.03157895
97/X	0.97	0.97979798	0.98979592	1	1.01041667	1.02105263
96/X	0.96	0.96969697	0.97959184	0.98969072	1	1.01052632
95/X	0.95	0.95959596	0.96938776	0.97938144	0.98958333	1
94/X	0.94	0.94949495	0.95918367	0.96907216	0.97916667	0.98947368
93/X	0.93	0.93939394	0.94897959	0.95876289	0.96875	0.97894737
92/X	0.92	0.92929293	0.93877551	0.94845361	0.95833333	0.96842105
91/X	0.91	0.91919192	0.92857143	0.93814433	0.94791667	0.95789474
90/X	0.9	0.90909091	0.91836735	0.92783505	0.9375	0.94736842
89/X	0.89	0.8989899	0.90816327	0.91752577	0.92708333	0.93684211
88/X	0.88	0.88888889	0.89795918	0.90721649	0.91666667	0.92631579
87/X	0.87	0.87878788	0.8877551	0.89690722	0.90625	0.91578947
86/X	0.86	0.86868687	0.87755102	0.88659794	0.89583333	0.90526316
85/X	0.85	0.85858586	0.86734694	0.87628866	0.88541667	0.89473684
84/X	0.84	0.84848485	0.85714286	0.86597938	0.875	0.88421053
83/X	0.83	0.83838384	0.84693878	0.8556701	0.86458333	0.87368421
82/X	0.82	0.82828283	0.83673469	0.84536082	0.85416667	0.86315789
81/X	0.81 *	0.81818182	0.82653061	0.83505155	0.84375	0.85263158
80/X	0.8	0.80808081	0.81632653	0.82474227	0.83333333	0.84210526
79/X	0.79	0.7979798	0.80612245	0.81443299	0.82291667	0.83157895
78/X	0.78	0.78787879	0.79591837	0.80412371	0.8125	0.82105263
77/X	0.77	0.77777778	0.78571429	0.79381443	0.80208333	0.81052632
76/X	0.76	0.76767677	0.7755102	0.78350515	0.79166667	0.8
75/X	0.75	0.75757576	0.76530612	0.77319588	0.78125	0.78947368
74/X	0.74	0.74747475	0.75510204	0.7628866	0.77083333	0.77894737
73/X	0.73	0.73737374	0.74489796	0.75257732	0.76041667	0.76842105
72/X	0.72	0.72727273	0.73469388	0.74226804	0.75	0.75789474
71/X	0.71	0.71717172	0.7244898	0.73195876	0.73958333	0.74736842
70/X	0.7	0.70707071	0.71428571	0.72164948	0.72916667	0.73684211

not required, because the inverse value of the required ratio can be calculated and the corresponding reciprocal ratio can then be used in the gear train. In a similar vein, the value of ratios such as *x*/50, *x*/49, *x*/48, and so on need not be calculated. These values are repeats of larger ratios already calculated, for example, 49/50 × 2/2 = 98/100. These two shortcuts reduce the number of calculations required to a quarter of what would be needed to produce a full table of gear ratios for up to 100 teeth.

ANALYSIS OF THE GEAR TRAIN

To begin this analysis, let us start with the ratios used in the gear train of Table 9. Most shafts contain a gear and pinion mounted on the same shaft.

Table 9. Gear Train Ratios

Shaft	No. of Teeth	Ratio	Ratio in Decimal	Rate of Rotation	1/Rate of Rotation
Hour	30			2 rev/day	0.5 day/rev
1st	60/36	30/60	0.5	1 rev/day	1 day/rev
2nd	48/16	36/48	0.75	0.75 rev/day	1.33333333 day/rev
3rd	64/12	16/64	0.25	0.1875 rev/day	5.33333333 day/rev
4th	72/6	12/72(or 1/6)	0.166666666	0.03125 rev/day	32 day/rev
5th	73/97	6/73	0.08219178	0.002568493151 rev/day	389.33333333 day/rev
6th	91	97/91	1.065934066	0.002737844347 rev/day	365.2508591 day/rev

The tooth count for both the gear and pinion are listed under the column labeled "No. of Teeth." For each gear listed, the first tooth count mates with the gear above, the second tooth count mates with the gear below.

Note the ratios between the hour, first, second, and third shafts. They are fractions that can be arrived at by division of whole numbers. In my search for a documented procedure of calculating orrery gearing, I have come across such a method, which is described in the text that follows.

The ratio between the hour shaft and the first shaft is 1/2, to produce a rate of 1 rev/day, which is used to drive the second shaft. Now identify the ratios for shafts 2 and 3 for a gear train that revolves once with 365.25 revolutions of the input shaft. Divide the rate by whole numbers as follows:

365.25 can be stated as 36525/100

5)36525(7305
36525

5)7305(1461
7305

3)1461(487
1461

These divisors can be combined (5 × 5 × 3 = 75) yielding 75/100 or 0.75 for a ratio of 3/4.

Continuing with the whole-number division we get:

4)487(121
484
3 the error is 3/4 tooth

6)121(202920
120
1 the error is 1/6 tooth

We now get the additional ratios of 1/4 and 1/6. At this point we have the first four ratios Rittenhouse used for the calendar gear train (1/2 for shaft 1, 3/4 for shaft 2, 1/4 for shaft 3, and 1/6 for shaft 4). Though not the most elegant approach, it is a method I have located in books of this era.

It is at this point that the process changes to what I would describe as the best-fit method. Any ratio errors resulting from the initial selections are accounted for in the final gear selections. I have not found this method of determining gear ratios described in any text. It is a method I have devised by conjecture as to how he might have arrived at the final two ratios of the train.

Starting at the first shaft, which is 1 rev/day, the ratio at this point is 3/4 × 1/4 × 1/6 = 3/96 or, in decimal terms, 0.03125.

The result now is a rotational rate of 235.25 rev/yr × 0.03125 = 11.4140625 rev/year for the forth shaft. The ratio required to complete the train is (1rev/yr)/(11.4140625 rev/yr) or 0.087611225.

The rules for the number of teeth on the remaining gears will be to select pinions of five leaves or more and gears with fewer than 100 teeth.

Trying a Five-Leaf Pinion

With a pinion of 5 and the required ratio of 0.087611225, the gear tooth count would be 5 × 1/0.087611225 = 57.07031262. Because we can only have whole teeth, we are required to use a gear of 57 teeth, and the ratio would then be 5/57, which is a ratio of 0.087719298, or 0.087719298/0.087611225 = 1.001233555. As perfect gearing would produce a result of 1.00000, the ratio error is 1.0000 − 1.001233555 = 0.001233555. To complete the gear train with this error would require a gear of 1/0.001233555 or 811 teeth.

Trying a Six-Leaf Pinion

With a pinion of 6 the tooth count would be 6 × 1/0.087611225 = 68.48437515. A ratio of 6/68 in decimal terms is 0.088235294, or 0.088235294/0.087611225 = 1.007123164. A ratio error of 0.007123164 would require a gear of 1/0.007123164 or 140 teeth to complete the gear train.

Trying an Eight-Leaf Pinion

With a pinion of 8 and the required ratio of 0.087611225 the tooth count would be 8 × 1/0.087611225 = 91.3125002. With a gear of 91 teeth the ratio would then be 8/91, which is 0.087912087, or 0.087912087/0.087611225 = 1.00343068. This is would result in a ratio error of 0.00343068 and would require a gear of 1/ 0.00343068 or 292 teeth.

From the preceding three calculations, a pinion of 6 leaves is the best choice because it will require the lowest tooth count on the mating gear. Having chosen a 6-tooth pinion, the challenge now is to identify the best final two ratios to complete the gear train. Having already determined that the ratio of 0.087611225 is required to complete the gear train, only ratios requiring fewer than 100 teeth will be considered for the best choices.

Trying different ratios in this vicinity using a 6-leaved driving pinion, the comparison exercise shown in Table 10 can be generated.

Because the required ratio to complete the gear train is 0.087611225, and the driving pinion is to have 6 leaves, we search a 100/100 ratio table and select a slightly larger ratio of 0.088235294 = 6/68 from the table as a starting point. We then divide the required ratio of 0.087611225 by the 6/xx trial ratio (0.087611225/0.088235294 = 0.992927218). We then repeat this with successive smaller ratios of 6/69, 6/70, 6/69 through 6/81 generating the column labeled "Gear Ratio." There is no set rule as to how many ratios to try; I have chosen 14 for this exercise.

The value 0.992927218 not being practical, continue with 1.007529096 (the value associated with 6/69) the closest value found in the ratio table is 1.01010101 with a corresponding ratio of 100/99. If the ratios of 6/69 and 100/99 are used to complete the gear train, the ratio error would be 0.002571914.

Table 10. Gear Ratio Exercise

Gear Ratio	The Ratio of 0.087611225 Divided by the Gear Ratio	Closest Table Ratio		Difference
6/68 = 0.088235294	0.992927218			
6/69 = 0.086956521	1.007529096	1.01010101	100/99	0.002571914
6/70 = 0.085714285	1.022130967	1.022222222	92/90	0.000091255
6/71 = 0.084507042	1.036732832	1.036585366	85/82	–0.000147466
6/72 = 0.08333333	1.051334742	1.051282051	82/78	–0.000052691
6/73 = 0.08219178*	1.065936581*	1.06593407*	97/91*	–0.000002511*
6/74 = 0.081081081	1.080538443	1.08045977	94/87	–0.000078673
6/75 = 0.08	1.095140313	1.095238095	92/84	0.000097783
6/76 = 0.078947368	1.109742189	1.109756098	91/82	0.000013909
6/77 = 0.077922077	1.124344067	1.125	99/88	0.000655593
6/78 = 0.076923076	1.138945939	1.138888889	82/72	–0.000057049
6/79 = 0.075949367	1.153547797	1.153846154	90/78	0.000298357
6/80 = 0.075	1.168149667	1.168831169	90/77	0.000681502
6/81 = 0.074074074	1.182751539	1.182926829	97/82	0.000175290

This procedure is now repeated for each of the remaining ratios of 6/xx in the table. When completed we find the smallest gearing error occurs when the train is completed with the ratios of 6/73 and 97/91 (marked with an asterisk), which are the same gearing ratios Rittenhouse used in the calendar gear train.

5

Equation-of-Time Indicator

THE ANALYSIS

The equation-of-time dial indicates the difference between 12 noon on a clock dial (mean solar time), and noon as indicated by a simple garden sundial (apparent solar time).

Mean solar time is the average length of all solar days in a year divided into 24 hours of equal length. These hours of equal length can be easily indicated on a mechanical clock. Noon, by clock indication, is the midpoint of the annual average day.

Noon by Apparent Solar Time on a sundial occurs when the Earth's meridian passes across the center of the Sun. When referenced to a clock, the occurrence of noon varies from the average throughout the year. There are two major causes for this variation.

In 1770 the reference one used to set the time on a clock was the sundial. So when using the garden sundial to determine the time, the equation-of-time correction factor was necessary for setting a clock to the proper time. If the garden sundial indicated 3:00 PM and the equation of time indicated the Sun to be 4 minutes slow on that day, the correct clock time was 3:04 PM.

The equation of time $(E = M - A)$ values are tabulated in Table 11 for each day of the year. When algebraically added to solar time, these correction values yield clock time. One should note that the equation-of-time dial on the clock is calibrated from 14 min. 45 sec. sun slower to 16 min. 15 sec. sun faster. Table 11 has a range of 14.4 min. (14 min. 24 sec.) "Sun Slower," to 16.4 min. (16 min. 24 sec.) "Sun Faster." The reason for this discrepancy is that the dial calibrations reflect the equation of time for the year 1770, whereas the table is for the year 2000. The difference is a drift that has occurred over 230 years.

One could use a table similar to Table 11 to set clock time using a sundial, or for convenience it could be an indication on the clock dial. For a dial indication, the common approach is to plot the values of the table in a polar form, with the radius representing the time correction and the angler position

Table 11. Equation-of-Time Correction Values

Day	Jan.	Feb.	Mar.	Apr.	May.	June	July	Aug.	Sept.	Oct.	Nov.	Dec.
1	3.6	13.7	12.5	4	-2.9	-2.4	3.6	6.2	0	-10.2	-16.3	-11
2	4	13.8	12.3	3.7	-3.1	-2.3	3.8	6.1	-0.3	-10.5	-16.4	-10.6
3	4.5	13.9	12.1	3.4	-3.2	-2.1	4	6.1	-0.6	-10.9	-16.4	-10.2
4	5	14	11.9	3.1	-3.3	-1.9	4.1	6	-0.9	-11.2	-16.4	-9.8
5	5.4	14.1	11.7	2.8	-3.4	-1.8	4.3	5.9	-1.2	-11.5	-16.3	-9.4
6	5.9	14.2	11.4	2.5	-3.5	-1.6	4.5	5.8	-1.6	-11.8	-16.3	-9
7	6.3	14.3	11.2	2.2	-3.5	-1.4	4.7	5.7	-1.9	-12	-16.3	-8.6
8	6.7	14.3	11	2	-3.6	-1.2	4.8	5.6	-2.2	-12.3	-16.2	-8.1
9	7.1	14.3	10.7	1.7	-3.7	-1	5	5.4	-2.6	-12.6	-16.1	-7.7
10	7.6	14.4	10.5	1.4	-3.7	-0.8	5.1	5.3	-2.9	-12.9	-16	-7.2
11	8	14.4	10.2	1.1	-3.7	-0.6	5.3	5.1	-3.3	-13.1	-15.9	-6.8
12	8.4	14.4	9.9	0.9	-3.8	-0.4	5.4	5	-3.6	-13.4	-15.8	-6.3
13	8.7	14.4	9.7	0.6	-3.8	-0.1	5.5	4.8	-4	-13.6	-15.7	-5.9
14	9.1	14.3	9.4	0.4	-3.8	0	5.6	4.6	-4.3	-13.9	-15.5	-5.4
15	9.5	14.3	9.1	0.1	-3.8	0.2	5.8	4.4	-4.7	-14.1	-15.4	-4.9
16	9.8	14.2	8.8	-0.1	-3.8	0.4	5.9	4.3	-5	-14.3	-15.2	-4.4
17	10.2	14.2	8.5	-0.4	-3.7	0.6	6	4	-5.4	-14.5	-15	-3.9
18	10.5	14.1	8.3	-0.6	-3.7	0.9	6	3.8	-5.7	-14.7	-14.8	-3.4
19	10.8	14	8	-0.8	-3.7	1.1	6.1	3.6	-6.1	-14.9	-14.6	-3
20	11.1	13.9	7.7	-1	-3.6	1.3	6.2	3.4	-6.5	-15.1	-14.4	-2.5
21	11.4	13.8	7.4	-1.2	-3.6	1.5	6.2	3.1	-6.8	-15.3	-14.1	-2
22	11.7	13.7	7.1	-1.4	-3.5	1.7	6.3	2.9	-7.2	-15.4	-13.9	-1.5
23	11.9	13.5	6.8	-1.6	-3.4	1.9	6.3	2.6	-7.5	-15.6	-13.6	-1
24	12.2	13.4	6.5	-1.8	-3.4	2.1	6.3	2.4	-7.9	-15.7	-13.3	-0.5
25	12.4	13.2	6.2	-2	-3.3	2.2	6.4	2.1	-8.2	-15.8	-13	0
26	12.6	13.1	5.8	-2.2	-3.1	2.4	6.4	1.8	-8.6	-15.9	-12.7	0.5
27	12.9	12.9	5.5	-2.4	-3.1	2.8	6.4	1.5	-8.9	-16	-12.4	1
28	13	12.7	5.2	-2.5	-2.9	3	6.3	1.3	-9.2	-16.1	-12.1	1.5
29	13.2		4.9	-2.7	-2.8	3.2	6.3	1	-9.6	-16.2	-11.7	2
30	13.4		4.6	-2.8	-2.7	3.4	6.3	0.7	-9.9	-16.3	-11.4	2.5
31	13.6		4.3		-2.6		6.3	0.4		-16.3		3

representing the day of the year. This produces a circular kidney-shaped graph as shown in Figure 25. This shape would be placed on a disk, cut out and mounted on a shaft that revolves once per year. A spring-loaded arm resting against the edge of the disc will follow the contour. The movement of the arm can then be used to position a pointer to indicate the corresponding sundial correction for each day.

The sign for the equation of time is dependent on whether one is converting Sun time to clock time or clock time to Sun time. It is positive during January, February, and March for converting Sun time to clock time, and would be negative for converting clock time to Sun time.

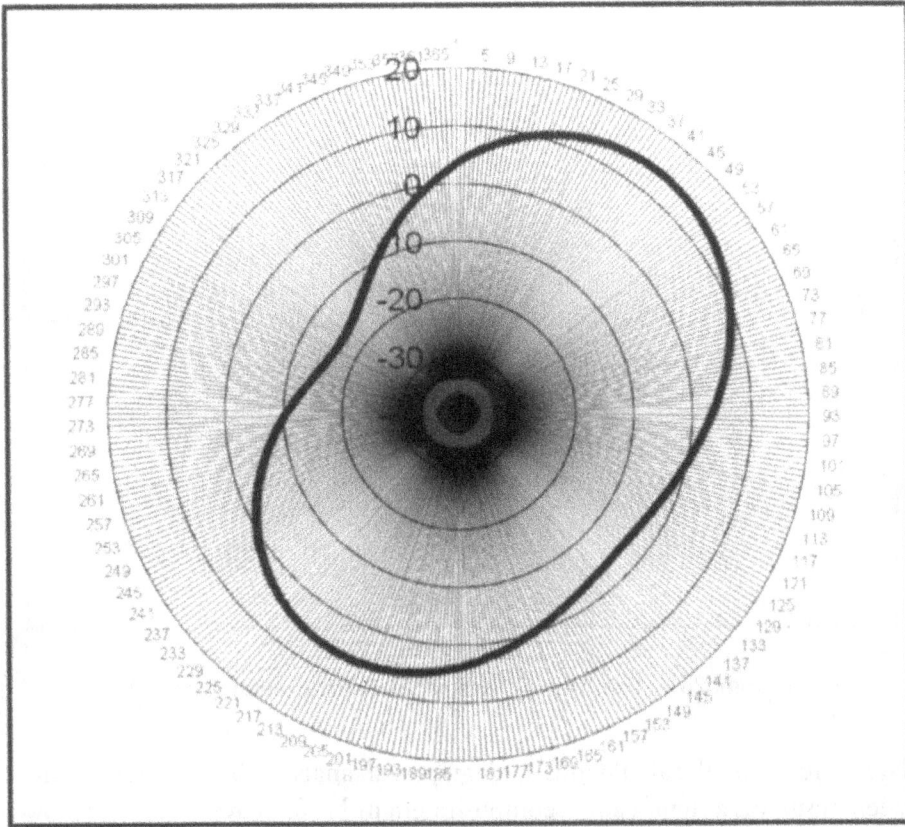

FIGURE 25.   Polar Plot of the Equation of Time

The time range of the radius used for the plot in Figure 25 was +20 min to −40 min. A smaller range could be used. However, as the negative end of the range is reduced, for example, such as from −40 min to −20 min, the depth and slope of the cusp increase considerably.

If the diameter of the disc is to be six inches, the radius would be three inches. If the three-inch radius is to represent fifty-five minutes (+15 min to −40 min), one minute is represented by a change of 0.055 inch in the length of the radius (3/55 = 0.0545). So to maintain an indication accuracy of one minute, the profile must be traced and filed to within 0.055 of an inch over its polar plot of the equation of time.

Rittenhouse, however, chose to take a different approach. He understood the underlying causes that result in the annual variations between clock and sundial time. So he chose to build a mechanical device that would replicate these deviations via gearing. I do not know of another example in which the equation of time has been implemented in this manner. Without computers and calculators, the only available means of analytical investigation in those

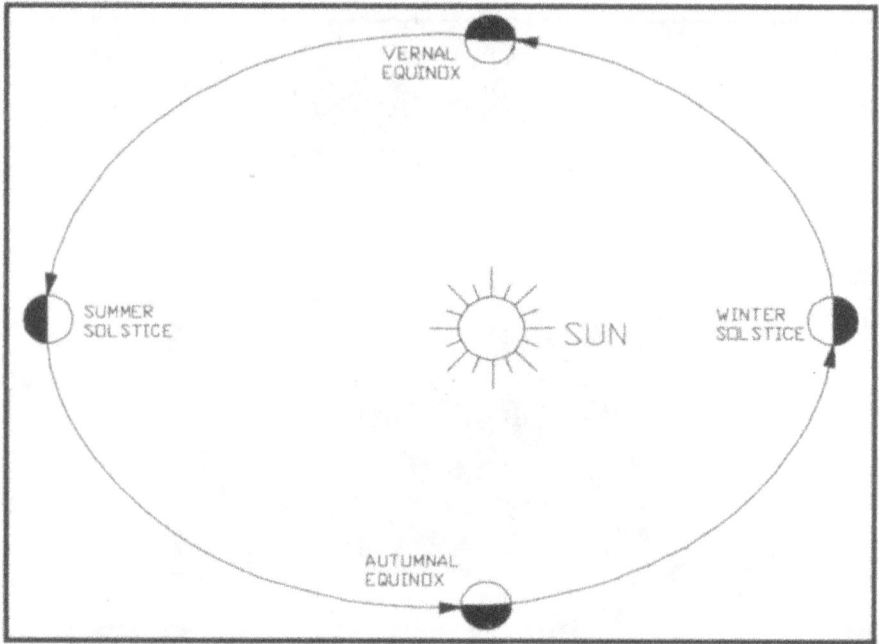

FIGURE 26.   The Annual Path of the Earth Around the Sun

days were manual calculations and graphical analysis. The design of this mechanism was a significant accomplishment in 1770. I have done the following analysis with the aid of a computer to gain insight to his design and approach.

One of the annual variations is caused by the progress of the Earth's motion in its annual orbit around the Sun. The path of the Earth is not entirely circular, nor is it exactly centered about the Sun. This is illustrated in Figure 26. Note that the orbital distance between the Earth and the Sun varies. The result is that for six months the velocity of the Earth increases, whereas for the remaining months it decreases.

Figure 27 illustrates the change in velocity for different positions in the Earth's orbit. The velocity varies such that for equal increments of time the area of the arcs swept are always equal. The velocity variation results in a contribution to the equation of time that is sinusoidal and has a repetition rate of one cycle per year. The cyclic amplitude in the occurrence of noon, because of this variation, is + and − 7.665 minutes. The zero crossing points occur at the beginning of January and the beginning of June. The negative maximum occurs at the beginning of April, whereas the positive maximum occurs at the beginning of October. The plot of this variation will be seen in Figure 29 as the wave that has a rate of one cycle per year.

The second variation is caused by the 23.5 degree tilt of the Earth's rotational axis, which is illustrated in Figure 28. This tilt produces a sinusoidal variation

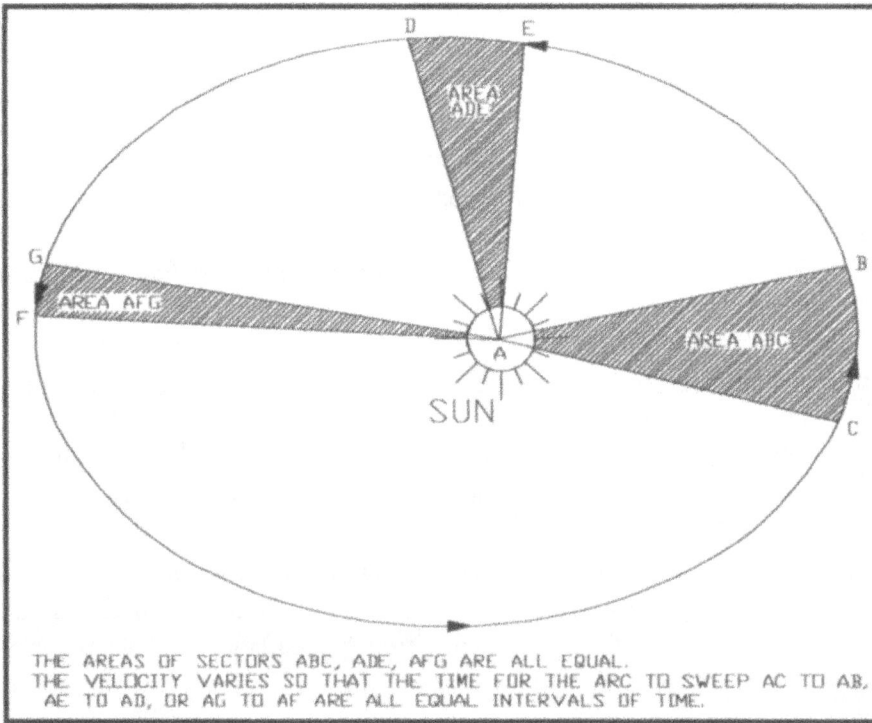

FIGURE 27. Variations in the Velocity of the Earth
for Different Positions in Its Orbit

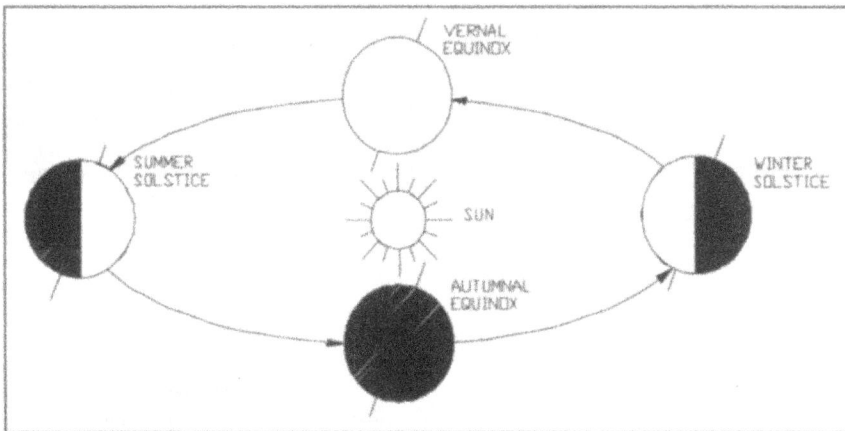

FIGURE 28. Seasonal Changes Are Caused by Tilt of Earth's Rotational Axis

with an amplitude of + and − 9.665 minutes at a rate of two cycles per year. The zero crossing points of this wave occur at the equinoxes (about March 21 and September 23) and the solstices (about December 21 and June 21). The negative maximum variations occur at the beginning of February and August, with the positive maximums occurring at the beginning of May and November. The waveform of this second, two-cycles-per-year variation, is also shown in Figure 29.

The zero crossings of the two waveforms do not coincide. There is a phase offset between the two wave forms. If the annual wave form for the eccentricity of the Earth's orbit is used as a reference, the zero crossing of the waveform resulting from the Earth's tilt is offset by 81 days from the zero crossing of that from the solar orbit.

The summation of the two waveforms shown in Figure 29 appears in Figure 30. Plotting the values from the equation-of-time table of Table 11 will produce a very similar graph.

In the real world, however, the actual variations are not truly sinusoidal. If we chose to use true sinusoidal waves for the two variations, incrementally sum them and then subtract the sum from the equation-of-time values of Figure 30, a computer analysis will reveal an error, which is illustrated in Figure 31.

To compensate for the error caused by using pure sinusoidal values, Rittenhouse uses a third wave to minimize this error. It is not an easy curve to minimize with a single trigonometric term. The one that works best is $0.31\text{Cos}(2B)$. This term is plotted in Figure 32 along with the resulting error when it is algebraically added to the first two sine waves. The magnitude of the error reduction when this term is included is 13.2 seconds, causing the maximum magnitude of the error to now be 52.8 sec (0.88 min × 60 sec/min = 52.8 sec).

A computer analysis indicates that these three terms produce the best sinusoidal simulation for the equation of time with the resulting equation being:

$$E = 7.665\text{Sin}(A) - 9.665\text{SIN}(2B) - 0.31\text{Cos}(2C)$$

The trigonometric angles A, B, and C can be expressed in terms of the annual day of the year:

$$A = [(360°/365.25)N]$$
$$B = [(360°/365.25)(N-81)]$$
$$C = [(360°/365.25)(N-173)]$$

Where N = Day of the year 1 through 365, the count beginning on January 1. The 360° of an annual revolution are divided into 365.25 increments (one increment for each day of the year). The phase relationship between A and B expressed in days, is equivalent to −81 days. The phase relationship between A and C is −173 days.

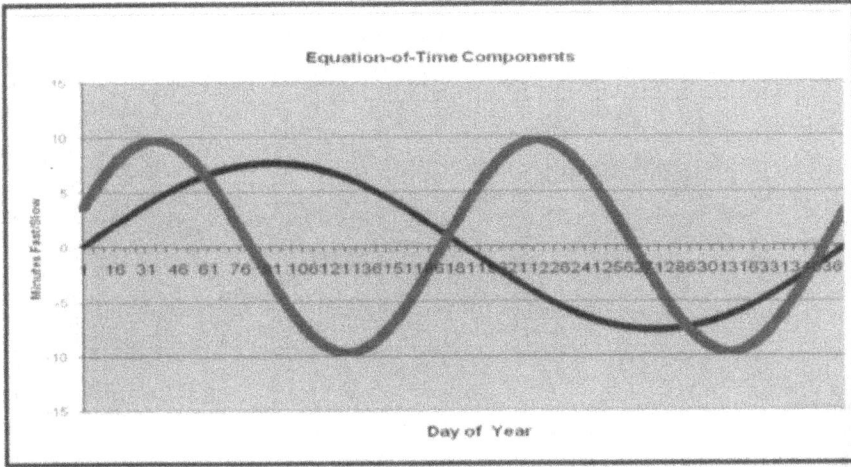

FIGURE 29.   The Two Variables That Form the Equation-of-Time Wave

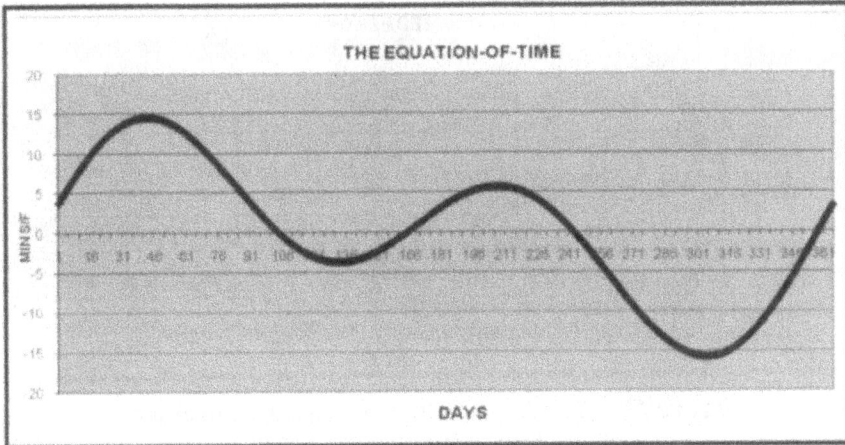

FIGURE 30.   The Equation of Time in Graphic Form

The results of this equation, overlaid with the values from the equation-of-time curve from Table 11, are illustrated in Figure 33. The difference between the curves, or error, appears as a low-amplitude curve along the horizontal axis.

THE MECHANISM

The mechanism that emulates this equation is shown in Figure 34, and is illustrated as a line drawing in Figure 35.

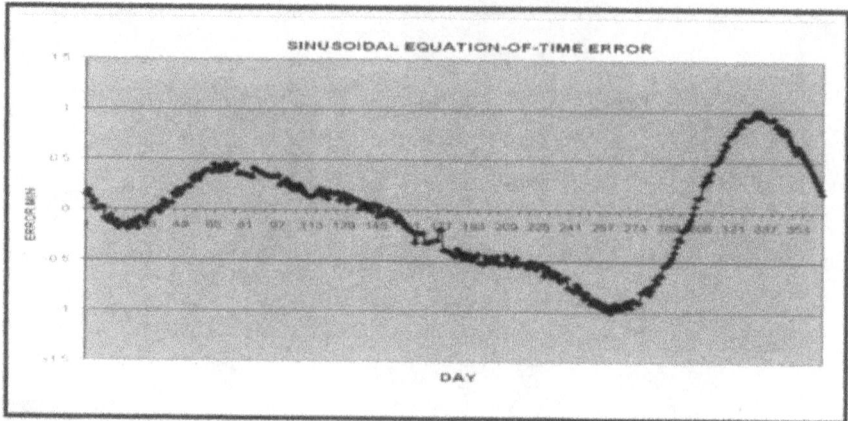

FIGURE 31. Difference Between the Values in the Equation-of-Time Table and an Emulation Using Two Sinusoidal Waveforms

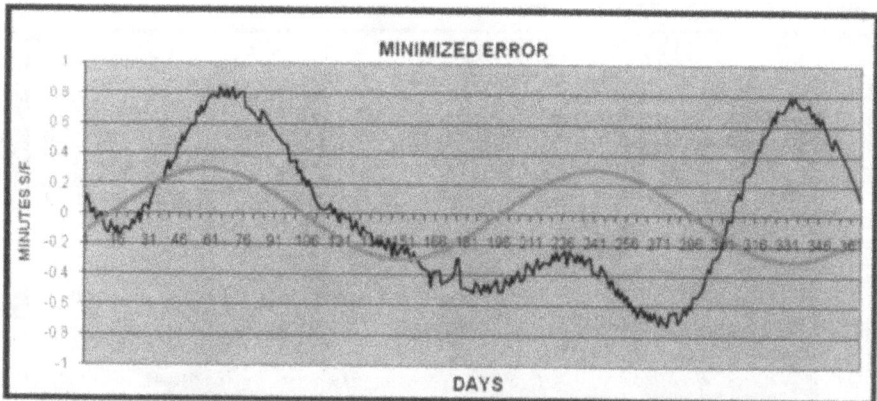

FIGURE 32. The Function 0.31cos(2D) (sine curve) and Reduced Error Amplitude (*jagged line*)

The E3 gear rotates once per year and carries the gear E6. The rotational axis of E6 is displaced from the axis of E3 by an amount that results in the gear contributing a variation that is equal to 7.665Sin(A). If the rotational axis of E6 were to engage the slot of E7, replacing the arm of E6 as the means of imparting motion to E7, the motion of the pivot alone would produce an amplitude of ±7.665 minutes variation on the dial indicator.

The arm on gear E6 generates the term 9.665Sin(2B). It makes two revolutions per year as E3 revolves once per year. The length of the arm on E6 is chosen to contribute a variation of ±9.665 minutes on the dial indicator as E6 rotates.

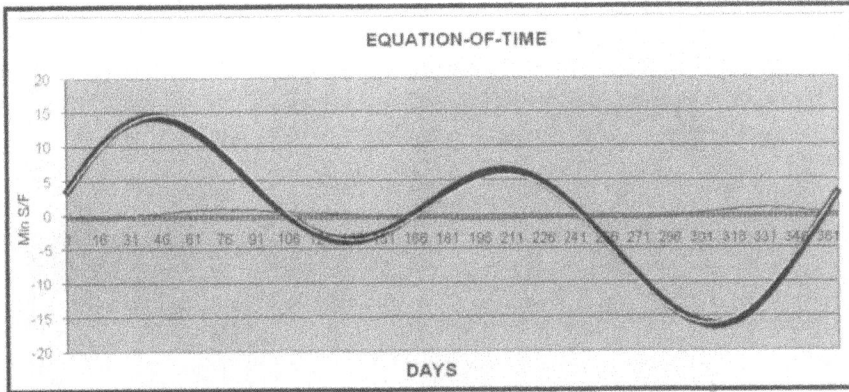

FIGURE 33.   Plot of the Equation of Time (*dark*) Against the Three-Term
Sinusoidal Equation (*white*) and the Resulting Error (*low-amplitude dark curve*)

The gear E4 generates the term 0.31Cos(2C) and also revolves two times per year. The length of the arm attached to gear E5 and the radius on which the E4 pin is placed produces a rocking motion to gear E5, such that it creates a variation of ±0.31 minutes on the dial indicator because of its rotational motion.

Phase relationships among the three terms of the equation are important in producing an accurate indication. The phase relationships among the three generating portions of the mechanism are established by the engagement of the gear teeth during assembly and the orientation of the arms during construction. To retain maximum accuracy, the original gear tooth engagements must be maintained.

The tooth counts of the gears in the mechanism are:

E3	72 teeth	7.665Sin(A)
E4	36 teeth	0.31Cos(2C)
E5	30 teeth	
E6	30 teeth	9.665Sin(2B).

The gear E3 has 72 teeth, which limits the smallest increment of engagement with E4 to increments of 5 degrees per tooth (360 deg/72 teeth). Gear E4 has 36 teeth, which limits its engagement to 10-degree increments. This implies that the angular position of the arm on E5 was chosen with great care, to obtain the required angular relationship between E4 and E6. Gears E5 and E6 both have 30 teeth, each providing engagement increments of 12 degrees per tooth.

Using the angles from the equation, at midnight on December 31 leap year +3, N equals zero. At this time the angle A in the equation is zero, [(360 / 365.25) × (0)]. If the vertical is used as the reference for 0 degrees, the pivot of gear E6 must be located on the vertical directly above the pivot of E5.

FIGURE 34. The Equation Mechanism

At this time the angle B is retarded, and behind the vertical in time by −79.8357 degrees, [(360/365.25) × (0 − 81)]. The angle C is also retarded, and behind by −170.5134 degrees, [(360 /365.25) × (0 − 173)].

To engage the gear E6 at an angle of −79.84 degrees, an angle of 72 degrees can be provided by engaging the gear six teeth late, each tooth providing 12 degrees. The remaining 7.84 degrees are accounted for in the positioning of

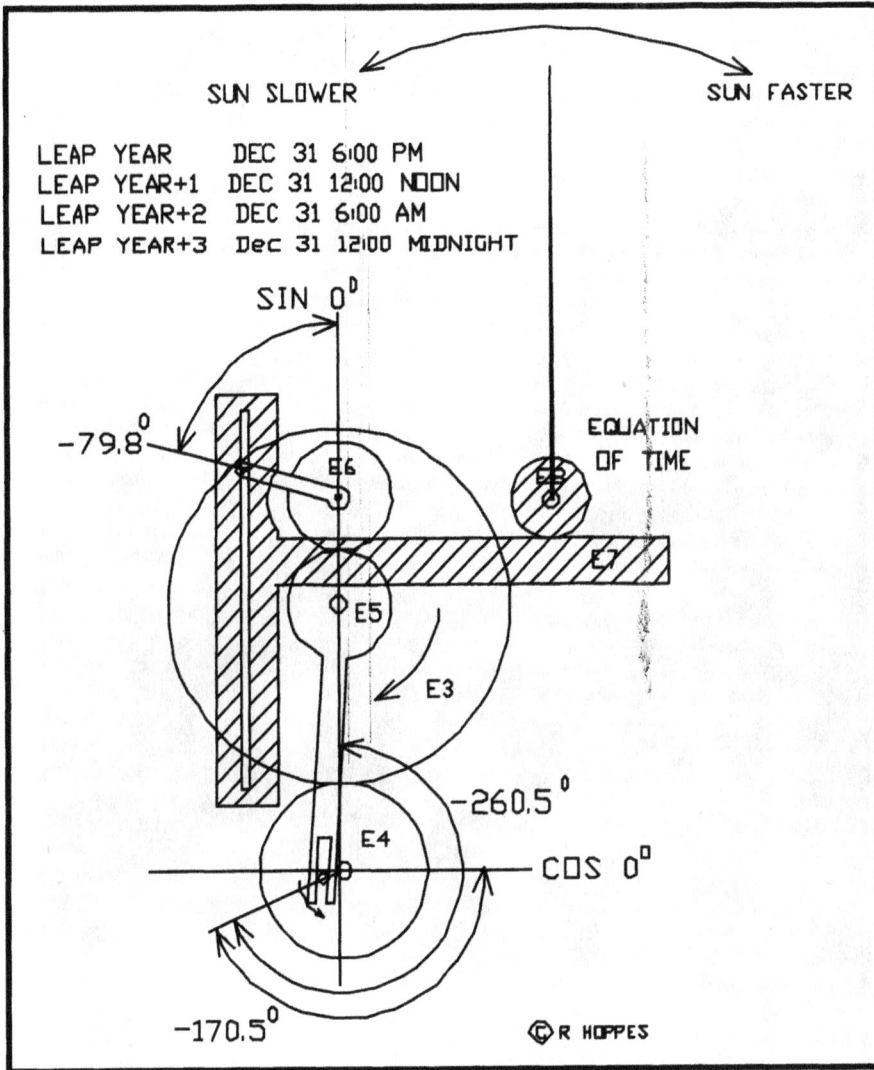

FIGURE 35. Gearing for the Equation-of-Time Indicator

the arm, when it was attached to the E6 gear. Figure 36 is a photograph of the E6 gear and attached arm. The E7 rack is also visible.

The E4 gear generates the 0.31Cos(2C) term by rocking the lever attached to E5. Because the term is a cosine, which is 90 degrees ahead of a sine, the horizontal plane becomes the reference for 0 degrees. Each tooth on this gear represents 12 degrees, so engaging it 14 teeth late provides 168 degrees of the cosine angle. However, one will discover it is more convenient to use the same vertical reference that was used for the previous two terms.

Consider that a sine advanced by 90 degrees can replace a cosine function. So if the angle −170.5134 degree is increased by an additional 90 degrees to

FIGURE 36.   The E6 Gear and Arm

−260.5134 degrees, the angle can be referenced to the same vertical that was used for the other sine functions. By retarding the engagement of the E4 gear by 21 teeth, we can obtain 252 degrees of the required negative angle (12 degrees per tooth). This leaves a remaining angle of 8.5134 degrees. This remaining 8.51 degrees is accounted for in the positional location of the pin on the E4 gear.

Figure 37 illustrates several progressive seasonal positions of the mechanism during the year. The views are numbered sequentially and illustrate the progression as E3 makes its annual rotation.

It is interesting to note the trace of the path created by the pin on the tip of the E6 arm over the interval of one year. This path is illustrated in Figure 38.

It is fascinating to note that if the larger upper loop were to be folded downward it forms a figure eight. This figure eight is representative of an analemma, as illustrated in Figure 39. An analemma represents the position of the Sun for each day of the year relative to a specific longitude at noon clock time every day. Globes frequently have an analemma illustrated on them in some area of the Pacific Ocean.

An examination of the equation-of-time dial in Figure 6 beginning at the top of the dial shows the indications are 0 to 14 minutes 45 seconds "Sun

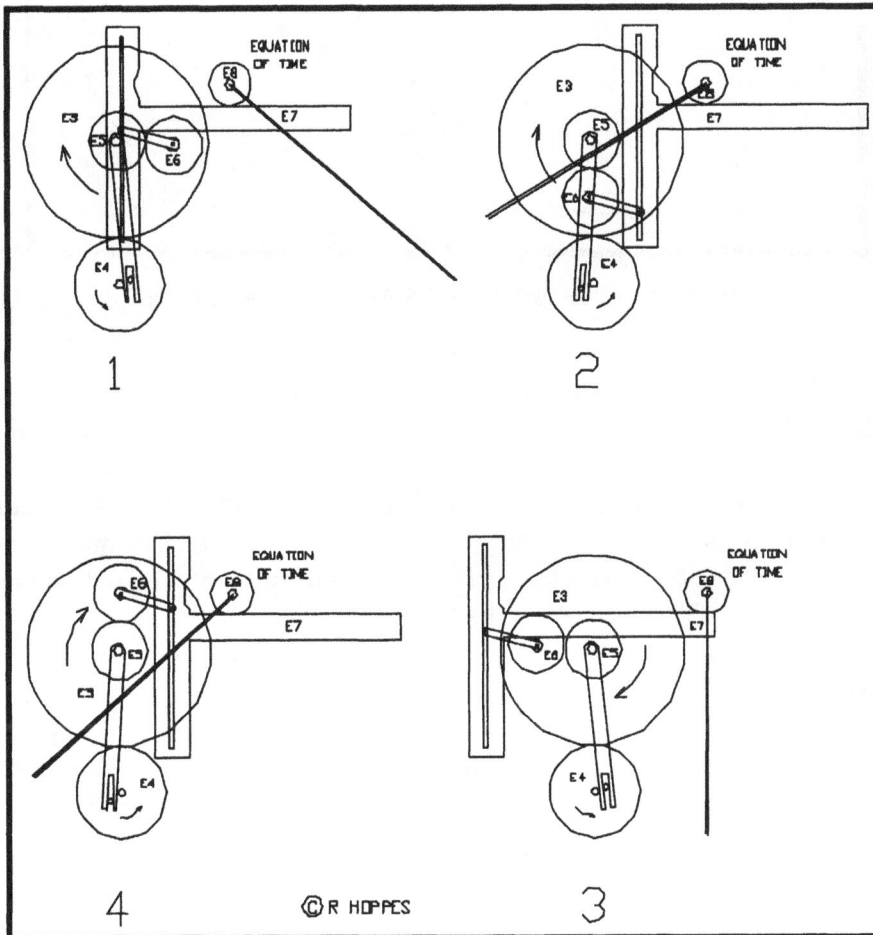

FIGURE 37.   Various Positions of the Mechanism During the Year

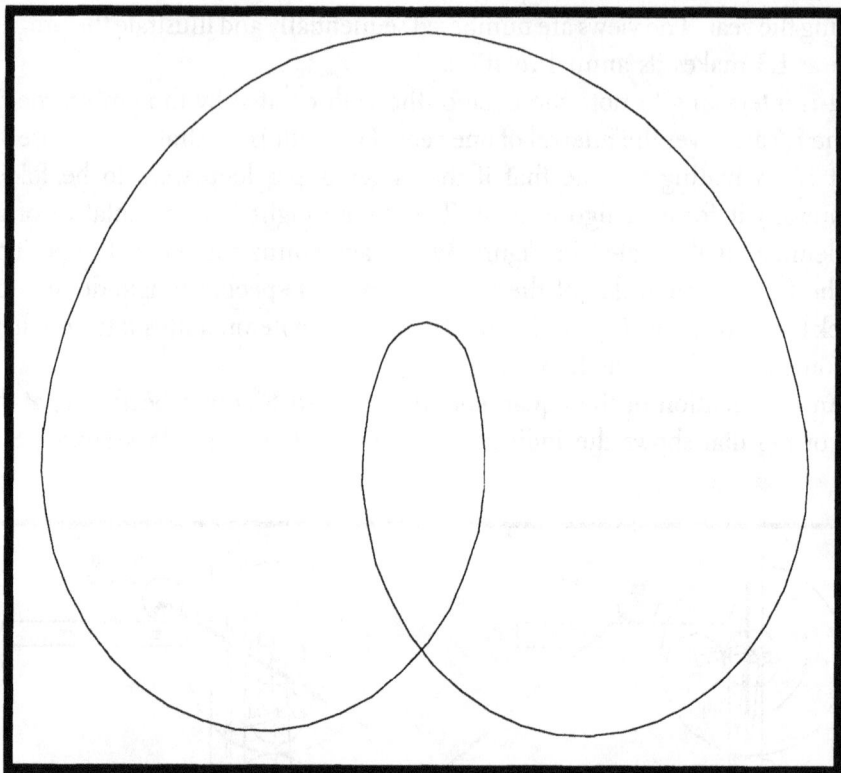

FIGURE 38. The Path of the E6 Arm Tip Over One Year

Slower" in the counterclockwise direction, to 16 min 15 seconds in the "Sun Faster" clockwise direction. There are an additional 10 seconds on each end, where the slow and fast indications meet. This makes the 360 degrees of the dial's circumference equal to the sum of (14 min 45 sec) + (16 min 15 sec) + 20 sec for a sum of 31 min 20 sec (31.333 min or 1,880 sec). The maximum indicator travel is 20 seconds less, so as to avoid confusion between indications of "Sun Fast' and "Sun Slow" at the times of maximum indication.

The arm lengths of the mechanism and the pitch diameter of gear E8 are dependent on each other. Assuming the pitch diameter (PD) is chosen first, the length of the toothed portion of E7 will be equal to one revolution of E8, or $\pi \times PD$ of gear E8. The equation-of-time mechanism has 40 active teeth to propel the pointer 360 degrees. Actually there are 40 to propel the pointer over 360 degrees plus 1 additional tooth for clearance to allow the fortieth tooth to engage.

The travel of E7 will be less by the ratio of the usable dial calibrations (1,860) and the total calibration increments (1,880), which extend over the 360 degrees of the dial. This makes the rotational travel of E8 = $\pi PD 1,860/1,880$.

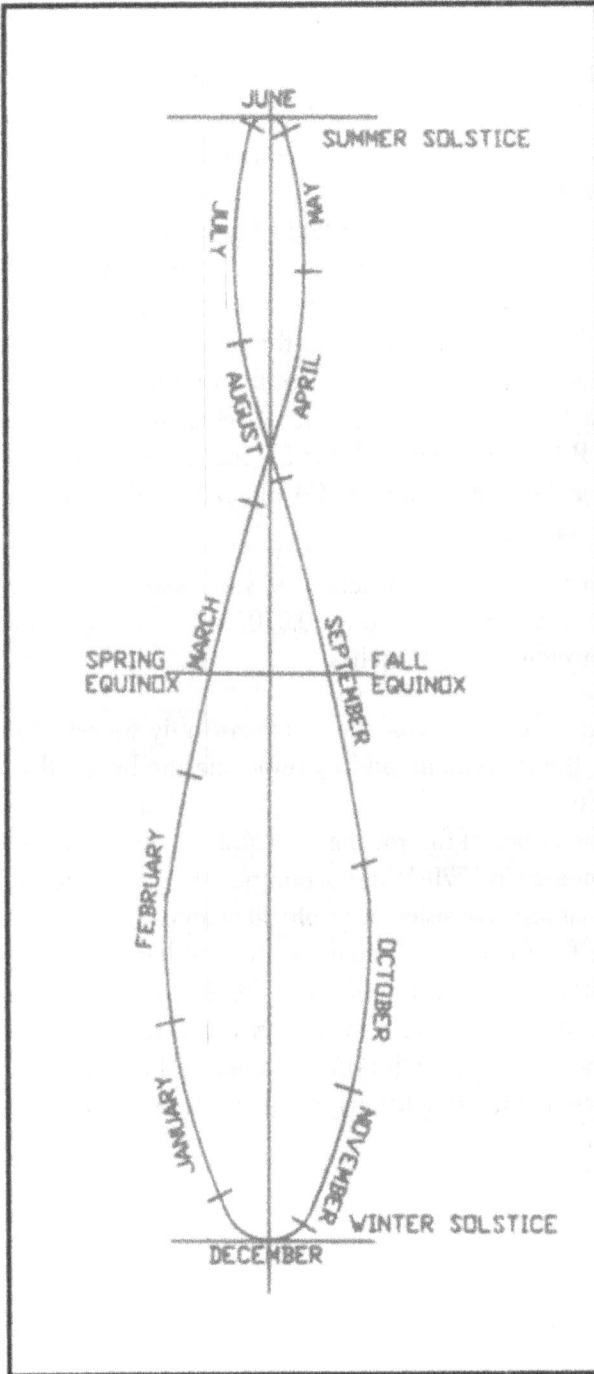

FIGURE 39. The Analemma

Because πPD = 31.333 minutes, 1 minute is equivalent to $\pi PD/31.333$. The length of the arm required on E6 becomes $9.665 \times \pi PD/31.3333$.

The distance between the pivots of E5 and E6 becomes $7.665 \times \pi PD/31.333$.

The ratios of E5 and E6 are 1:1, so both have the same diameters and radii. The distance between their pivots then equals the pitch diameters of the gears ($7.665 \times \pi PD/31.333$).

We know that a 90-degree revolution of gear E6 represents 9.665 minutes on the indication dial. To get E5 to impart a variation of 0.31 minutes will require a rotational action of 0.31min/(9.665 min/90deg) = 2.8867 deg. Because the ratio between E3 and E4 is 1 to 2, the radii also have a ratio of 1:2. The spacing between the pivots is the sum of their radii, and it is also the length of the arm attached to E5 when the pin of E4 is aligned horizontally with its pivot. For the Rittenhouse clock this is 2.2 inches. The angle of the variation can be stated as Sin 2.8867 deg = (E4 pin radius/ E5 arm length). Solving for the E4 pin radius we get:

> E4 pin radius = E5 arm length × Sin 2.8867
> E4 pin radius = 2.2 inches × 0.0503611
> E4 pin radius = 0.111inches

The angular relationships of the arms can only be set while the dial is removed from the movement, and by removing the bridge that retains gears E3, E5, and E6.

I consider the design of this mechanism quite an accomplishment considering it was engineered in 1770. Without computers or calculators the methods of evaluation at that time consisted of graphical analysis and manual calculations.

By the time I completed this analysis, the dial had already been mounted back on the movement and I could not obtain exact measurements. Trying to compare the dimensions from my analysis against those of the mechanism in the clock was encouraging, however. As best as I could tell without proper access, the calculated dimensions compare favorably with those of the clock's mechanism.

6

The Strike Operation

FIGURE 40 IS A PHOTOGRAPH of the strike components located on the front plate of the movement. The clock strikes hours and quarters on two bells. Hours are struck on the larger bell, whereas quarters are struck on the smaller one. The strike mechanism and the manner by which it operates are unique and different from other tall-case clocks. The hour rack and snail are shown in Figure 41, whereas the barely visible quarter snail appears in Figure 42.

All strike sequences begin with the strike of the hour followed by the striking of the quarters. The bells have individual hammers both of which are operated by the same lift lever, pinwheel, and gear train. The number of strikes on each bell is controlled via two racks. One rack counts the number of strikes for the hours, which then continues to be advanced during the quarter strikes, a second rack then counts the number of quarters and is responsible for ending the strike sequence. The method by which the sequence operates is intriguing.

THE MANUAL REPEAT AND WARNING PHASE

A strike sequence begins with a warning phase. Figure 43 illustrates the components involved in the warning. A wheel driven by the minute shaft rotates once per hour and, having four pins in its rim, causes the warning for each of the hour and quarter operations. The trip pins lift the warning lever in preparation for the strike, and releases it on the hour or quarters as selected by the setting on the dial. The hour pin is longer and protrudes out farther than the quarter-hour pins. By moving the warning lever closer to the front plate, striking will occur on every quarter because each pin will cause a trip. By moving it slightly farther out, striking will occur only on the hour because only the long pin will cause a trip, and moving it still slightly farther means

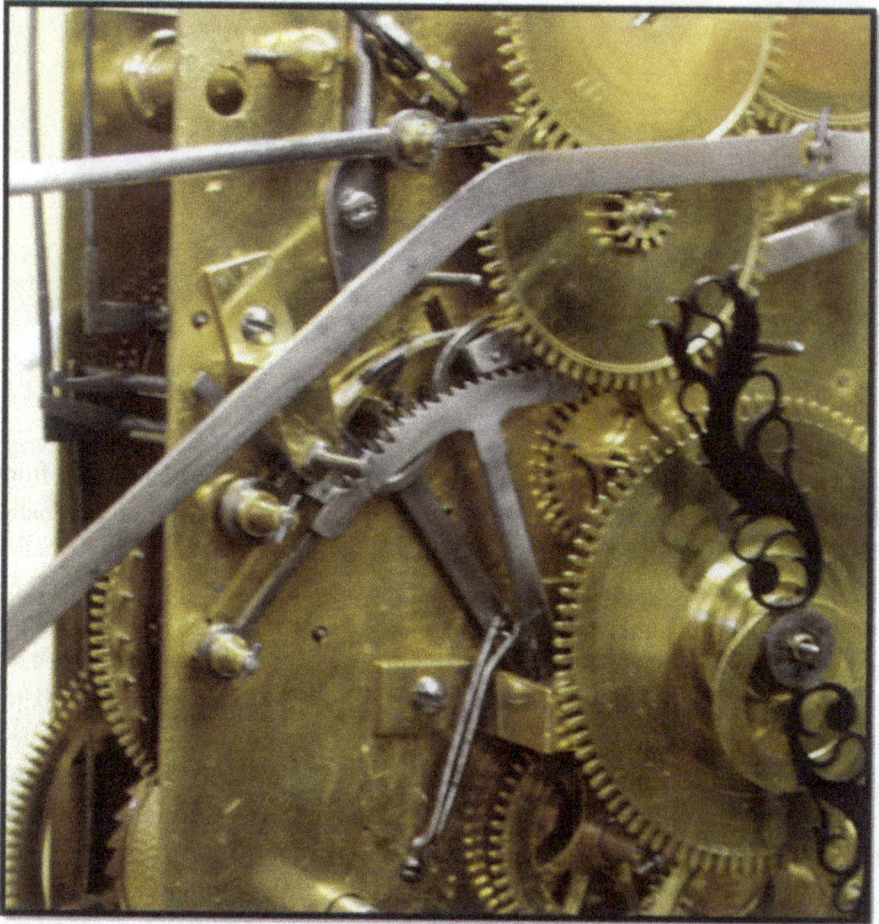

FIGURE 40.   The Strike Mechanism

no strikes will occur. This is similar to the operation of the music trip selection in Figures 66 and 67 in the description of the music train operation.

The last strike sequence may be repeated manually by lifting the repeat lever and releasing it. Although the repeat lever may be raised at any time, a strike may not begin if a strike warning operation is in progress. A repeat will not occur because the warning lever will not fall until the trip has been completed.

During a warning, when the lever is raised, the racks fall to the left (counter-clockwise) and the strike sequence is ready to begin. However, an extension on the left-hand end of the warning lever protrudes through an opening in the front plate, preventing the train from running as long as the lever is lifted. The protrusion prevents the train from running because it moves into the path of a pin placed in the rim on the wheel that drives the strike fly. The wheel is drawn with dashed lines to illustrate that it is located between the front and

FIGURE 41.   The Hour Snail and Both Racks

FIGURE 42.   The Quarter-Hour Snail

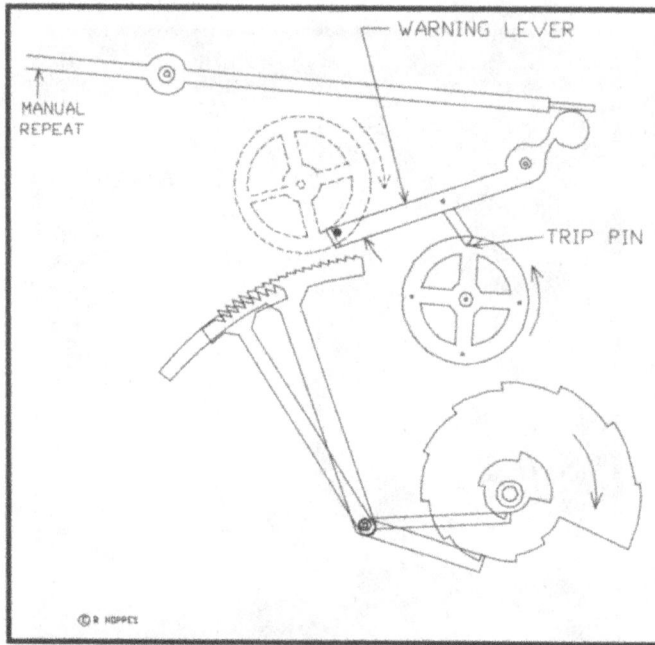

FIGURE 43.   The Manual Repeat and Warning Phase

rear plates. The strike train is arrested as long as the warning lever is in the raised position.

When the racks drop, their travel is limited by their tails, each of which contacts a step on their corresponding snail. The height of the step under each rack tail determines how many teeth are required to be gathered during the strike.

COUNTING THE STRIKES

Figures 44 and 45 illustrate the hour and quarter racks in greater detail. Note that the upper edge of the hour rack in Figure 44 has three distinct sections. Starting on the right-hand end of the sector and progressing toward the left, there are fourteen shallow teeth along the top edge, followed by four teeth that are cut approximately twice as deep as the shallow teeth. These deeper teeth are then followed by a smooth tail on the left-hand end. The surface of this smooth tail is located approximately at the same height as the bottom of the shallow teeth. The hour snail is mounted on the hour gear and rotates clockwise once in 12 hours. The snail is shown in the one o'clock position.

Figure 45 illustrates the quarter rack. Its tail and the quarter-hour snail against which it falls determine the number of quarters that will be struck.

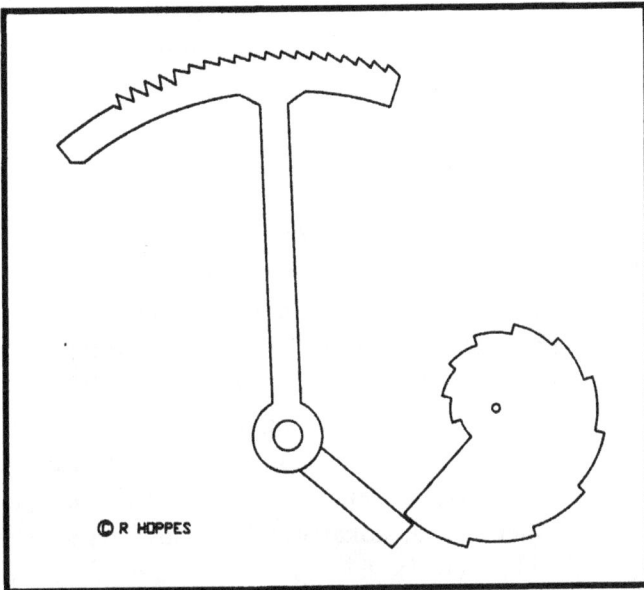

FIGURE 44. Hour Rack and Snail

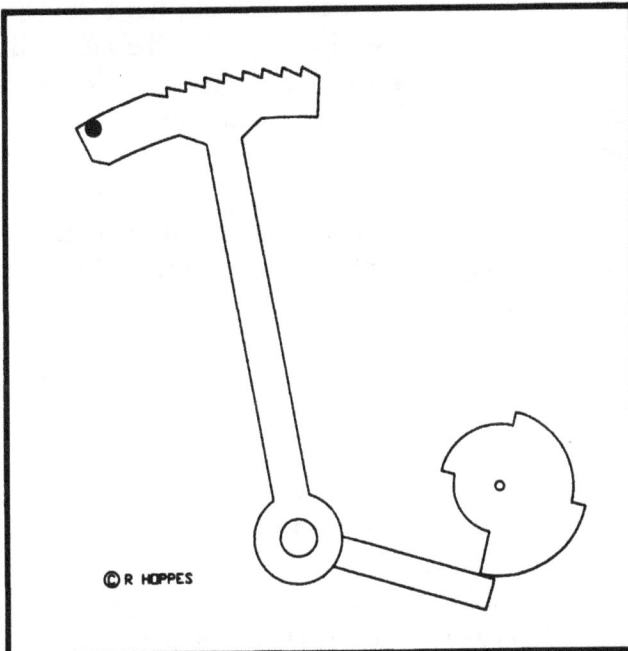

FIGURE 45. Quarter Rack and Snail

The quarter rack has eight teeth with a smooth section on the left-hand end. A pin protruding from the rear of the left end is used to terminate the strike sequence. The snail for the quarter rack is mounted on the minute shaft and rotates clockwise once per hour. The figure illustrates the position for the quarter strike sequence of 1, 2, 3, 4 or 0, 1, 2, 3 depending on the relative angle of the tail to the rack.

Figure 46 illustrates the hour rack, the gathering pallet (which lifts each rack one tooth for each strike), the rack hook (which engages the rack teeth and retains the rack position between gathers), and the shoe that rides on top of the rack teeth. The figure portrays the rack position prior to the last hour strike, when one shallow tooth is yet to be gathered.

When the hour rack falls on the warning, there will be one shallow tooth to the left of the rack hook for each hour that is to be struck. This is the starting position for the strike sequence. During the sequence the hour rack counts the total number of hour strikes and then continues to advance as the quarters are struck. This advance takes place on the teeth that are between the rack hook and the shoe. The gathering pallet rotates counterclockwise and advances the rack to the right one tooth for each strike.

The rack hook engages each tooth of the hour rack as it is advanced, preventing it from falling back to the starting position after it has been released by the gathering pallet.

One should notice where the shoe is located and the distance relationship between it and the tooth of the rack hook. This distance corresponds to a span of four teeth on the rack. Because of this spacing, the shoe will ride on the tips of the teeth as long as the rack hook is engaging shallow teeth. However, when the hook encounters the first deep tooth, the shoe will have dropped onto the smooth portion of the rack tail.

Once the shoe has descended to the lower position, hammer strikes will occur on the smaller bell. Although the gathering of the hour rack will continue as the quarters are struck, the number of quarter strikes is determined by the number of teeth to be gathered on the quarter rack. Striking continues until the pin on the tail of the quarter rack terminates the strike sequence.

Figure 47 illustrates the rack in position for the final quarter strike. The gathering pallet gathers both the hour and quarter rack one tooth for each quarter strike. When the last quarter strike has been completed, the pin on the quarter rack will have moved into the path of the gathering pallet tail, halting further rotation and ending the strike, as illustrated in Figure 48.

The joint action and interaction of the shoe, racks, and levers during the strike will now be described. It begins with the warning phase, which releases the gathering pallet and presets the racks to the proper positions for the strike. To allow the racks to fall during the warning, both the rack hook and the shoe must be lifted clear of the teeth on both racks.

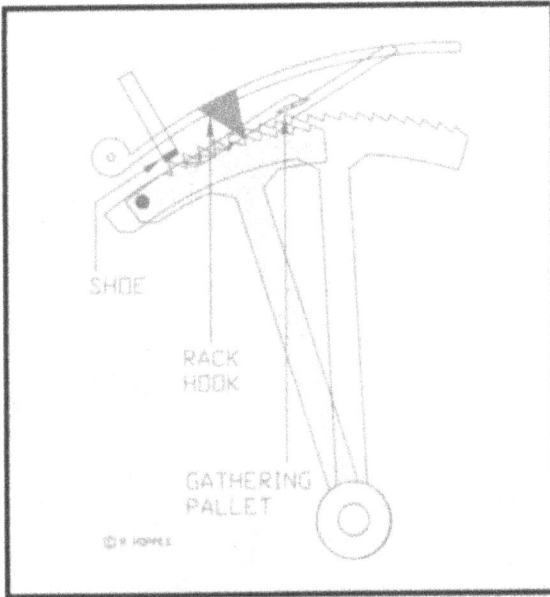

FIGURE 46. Hour Rack Position for an Hour Strike of One

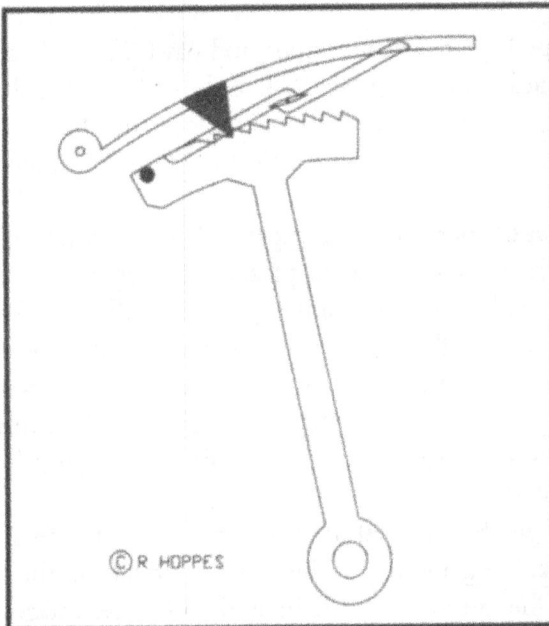

FIGURE 47. Quarter Rack Position for a Strike of One Quarter

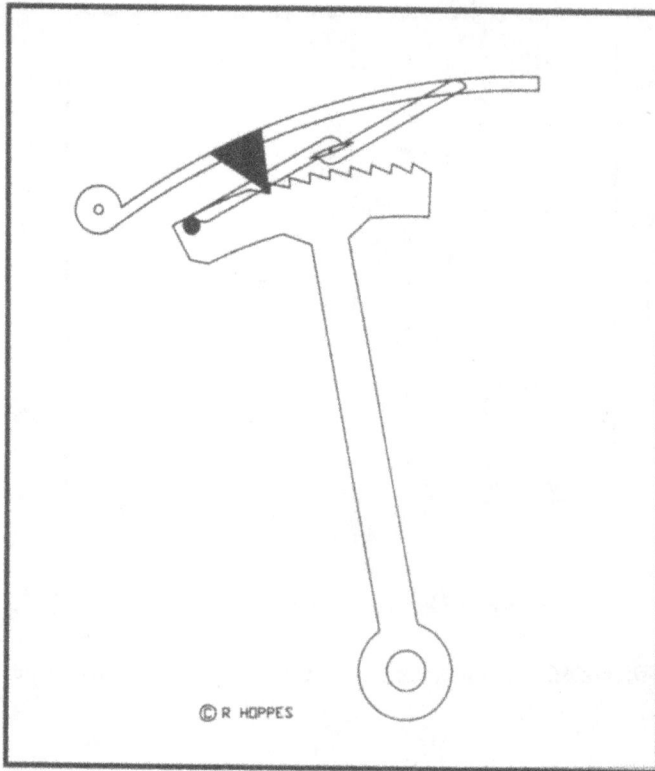

© R HOPPES

FIGURE 48.   Quarter Rack Position After Completing the Quarter Strike

Figure 49 depicts a photo of the shoe and pawl lift levers, as well as a line drawing of the racks, shoe, and levers. The shoe lift lever is shown shaded in the line drawing of Figure 49. The warning lever lifts both the shoe lift and rack hook levers, by applying upward lift to both as illustrated by the arrow in the figure.

The quarter and hour racks are separated by an ample distance, which allows for the shoe lift lever to be placed between them. Because of the separation of the racks, the leaves of the gathering pallet are extended (lengthened toward the dial) so they can engage the teeth of both racks for gathering. A downward curve placed in the shoe lift lever ensures there is sufficient clearance between the lever and the gathering pallet at all times, preventing interference between them. Upon release of the warning lever, the strike sequence begins.

The stages of greatest interest are the gathering of the final shallow tooth on the hour rack, the gathering of the first deep tooth on the hour rack, and the gathering of the last tooth on the quarter rack. These stages are illustrated in Figures 50, 51, and 52.

Figure 50 illustrates the hour rack in position to carry out the final hour strike. It is the final hour strike because the rack hook is engaging the last

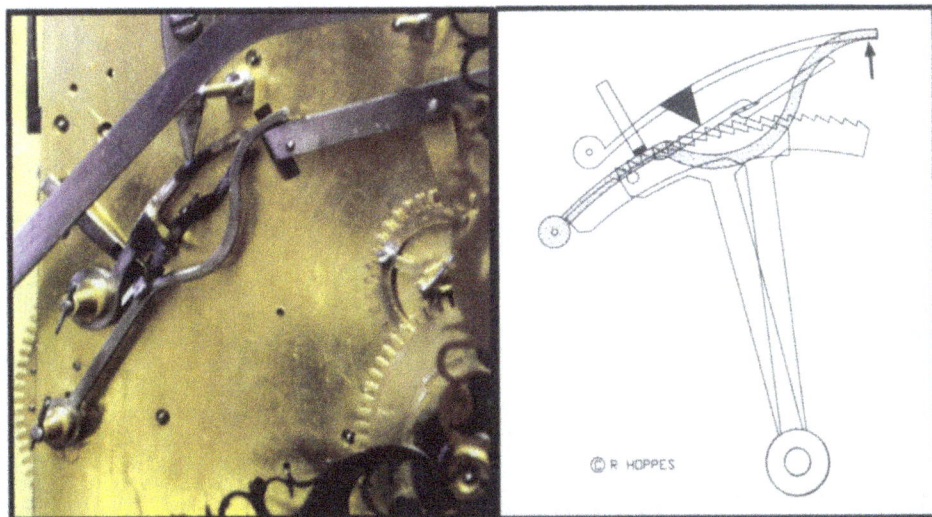

FIGURE 49. Warning Lift of Shoe and Pawl

shallow tooth on the rack, and the shoe will fall when the gathering and strike occur. Note that during hour strikes the rack hook engages only shallow teeth. As long as the shallow teeth are being engaged, the rack hook cannot descend low enough to prevent the quarter rack from falling back to the position established by the quarter snail. After performing the final hour strike, the racks and the hook will be as shown in Figure 51.

The things to note in Figure 51 are that the shoe has descended onto the smooth portion of the hour-rack tail. Because the shoe has descended, hammer strikes will now occur on the smaller bell. Also, the rack hook has now encountered the first deep tooth of the hour rack. When the rack hook encounters the deeper teeth it descends slightly lower and now also retains the teeth of the quarter rack, which prevents it from falling back to the position established by the quarter snail. As they are gathered, the positions of both racks will now be retained by the hook. The number of quarter strikes that will be heard is determined by the number of teeth still to be gathered on the quarter rack. The important detail to note is that after the last hour has been struck, the quarter snail is now retained by the rack hook, with its tail elevated above the quarter snail by the space of one tooth of the quarter rack.

Striking of the quarters will now occur, and gathering of both the hour and quarter racks will continue until the quarter rack has been gathered, as illustrated in Figure 52. Once the shoe has fallen onto the smooth tail of the hour rack, the gathering of hour-rack teeth that follows will correspond to the number of teeth gathered on the quarter rack. However, the gathering of the hour rack has no further influence in the sequence.

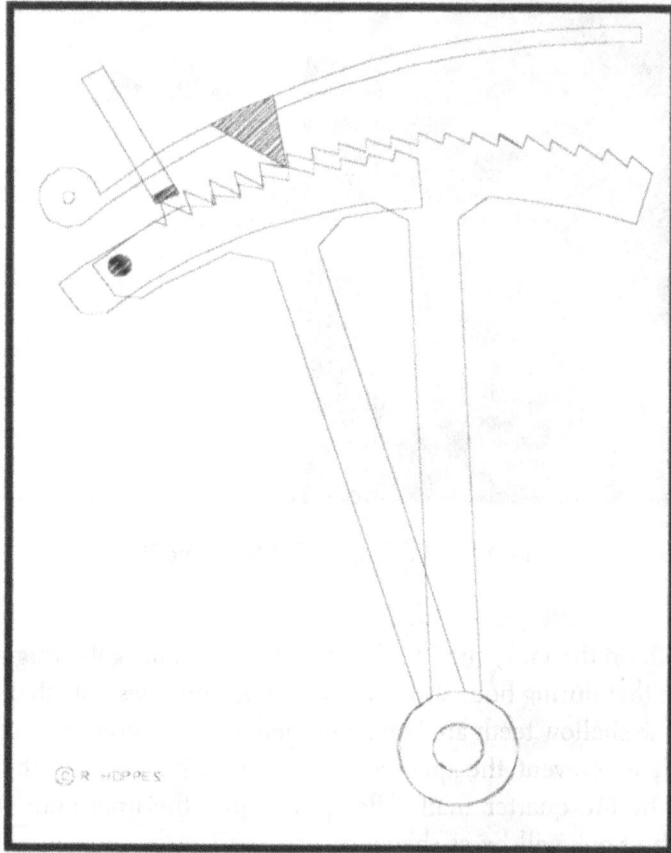

FIGURE 50. Rack Hook Ready to Tally the Final Hour Strike

When the teeth on the quarter rack have been gathered, as illustrated in Figure 52, the strike sequence will be completed. The strike sequence has been terminated, because the pin on the tail of the quarter rack has moved into the path of the gathering pallet tail, halting further rotation of the gathering pallet.

One should be aware that when the last hour strike has been completed in Figure 51, the tail of the quarter rack will no longer be resting on the quarter snail. Because the rack hook is now engaging a deeper tooth on the hour rack, the rack hook no longer allows the quarter rack to fall back onto the quarter snail. At this point the quarter rack will be retained at a position of one tooth above the step of snail. Or to state it slightly differently, prior to the rack hook engaging deeper teeth on the hour rack, the number of quarter-rack teeth exposed to be gathered was one plus the number of required quarter strikes. After the last hour strike, the number of teeth remaining to be gathered on the quarter rack will be equal to the number of quarters to be struck.

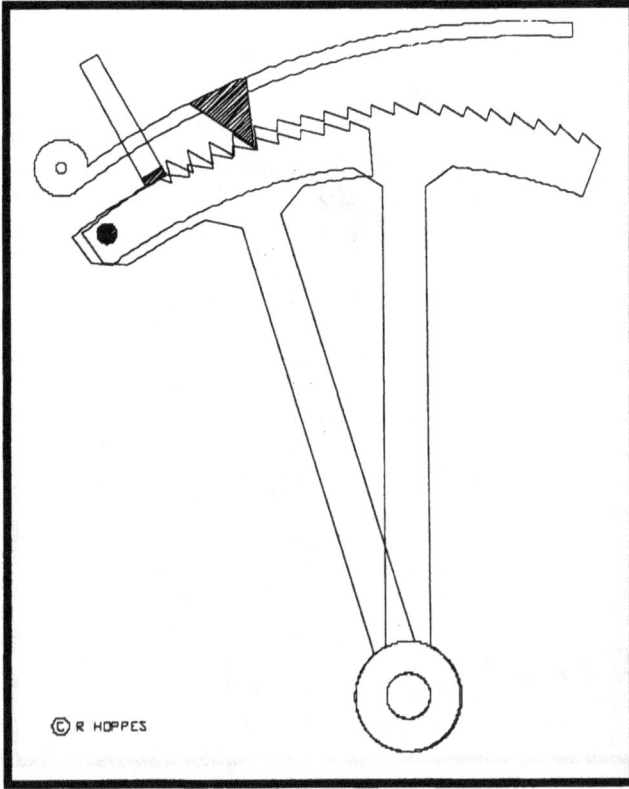

FIGURE 51.   Rack Hook Ready to Tally First-Quarter Strike

CONTROLLING THE BELL HAMMERS

A photo of the bell hammers and intermediate shaft are shown in Figure 53, with line drawings of the hammers and shafts in Figures 55 and 56.

Figure 54 is a photograph of the shoe in the lower position for quarter striking.

The control of which hammer to be lifted for the strike is performed by an intermediate shaft. This shaft has long extended pivots and an ample amount of end play. This gives it two degrees of freedom. Rotation of the shaft provides hammer lifting for strike, whereas end-to-end positional shifting allows for selection of which hammer is to be lifted. The hour hammer is selected when the shaft is fully positioned against the back plate, as illustrated in Figure 55. The active components are shaded in the figure.

When the shoe is no longer supported by the tips of the rack teeth, it drops onto the smooth tail portion of the rack. When it drops, the leaf spring exerts forward pressure on the rear pivot, shifting the shaft forward. In the forward position, rotation of the shaft actuates the quarter-hour hammer as illustrated in Figure 56.

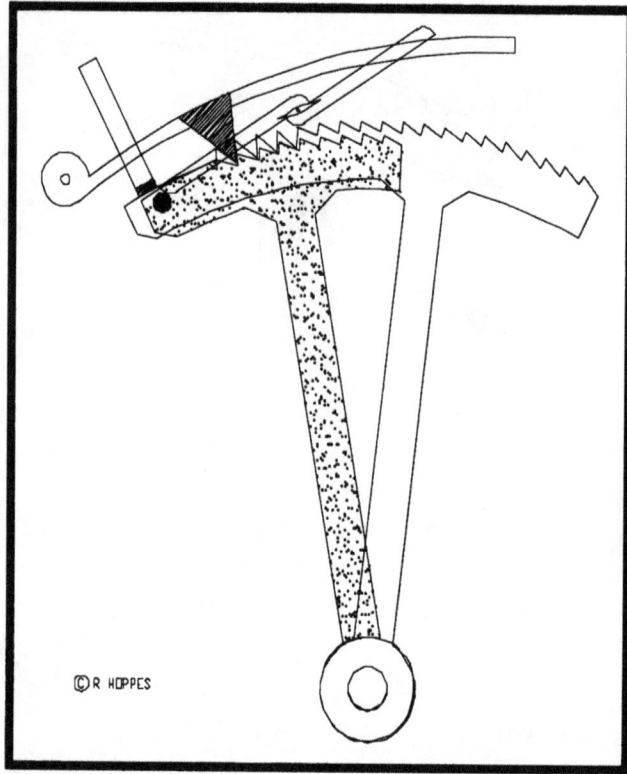

FIGURE 52.   Rack Hook After Final Quarter Strike

Figure 57 illustrates the hammer lift arms, the intermediate shaft, and the strike wheel lift pins, all viewed from the forward position. The shaded portion is the intermediate shaft, which shifts front to back, selecting which hammer is to be lifted.

This technique of front-to-rear shaft shifting to perform selection of a function is shown in Figures 55 and 56, and is used three times in this movement. First, for the selection of the hour and quarter-hour hammer strikes; second, for the selection of the strike interval (Silent, Hourly, Quarterly); and third, for the tune-play interval (Silent, 2 Hr, 1 hr, 1/2 Hr, 1/4 Hr). The individual leaf springs for these three shaft-shifting functions are visible in the rear view of the movement shown in the right- hand portion of Figure 59 in the following section. One is to the right, one to the left, and one in the center of the rear movement plate.

The strike, as it is presently, has been altered from its original design. It is the opinion of several others and myself that originally the clock struck hours only on the hour, without any quarter strikes. At the quarter hours, it struck the last hour followed by a corresponding number for the quarter.

FIGURE 53.   The Hammers

The current strike, of four quarters on the hour, is the result of the quarter snail being altered and its engagement with the gear train being retarded by a quarter of a turn. This, in conjunction with the arm of the quarter rack, which falls onto the quarter snail, is not in the proper relationship to the

FIGURE 54.   Shoe in Lower Position

teeth of its rack. These alterations result in producing a four-quarter strike on the hour.

Another reason for the opinion that the four-quarter strike is not original to the clock's design, is that when four quarters are struck on the hour the quarters are usually struck as a prelude to the striking of the hours, such as in the Austrian strike, Westminster, and so on. An additional reason is that if four quarters are not struck, the fall of the hour weight will power the strike train for 32 days, the same interval of time required for the time train weight to fall.

This is a significant point, because if the strike train weight descends in 18 days and is not rewound, the strike mechanism can become dormant in the midst of a strike, which is likely to cause the clock to stop. These reasons, along with the alterations to the quarter snail, suggest the strike has been modified.

The original return springs of the hour and quarter racks are missing. Figure 58 provides some insight into the form of the original spring. The view is less than desirable, but it appears to be the only existing record. In this photo the springs have already been bent and altered from the original configuration.

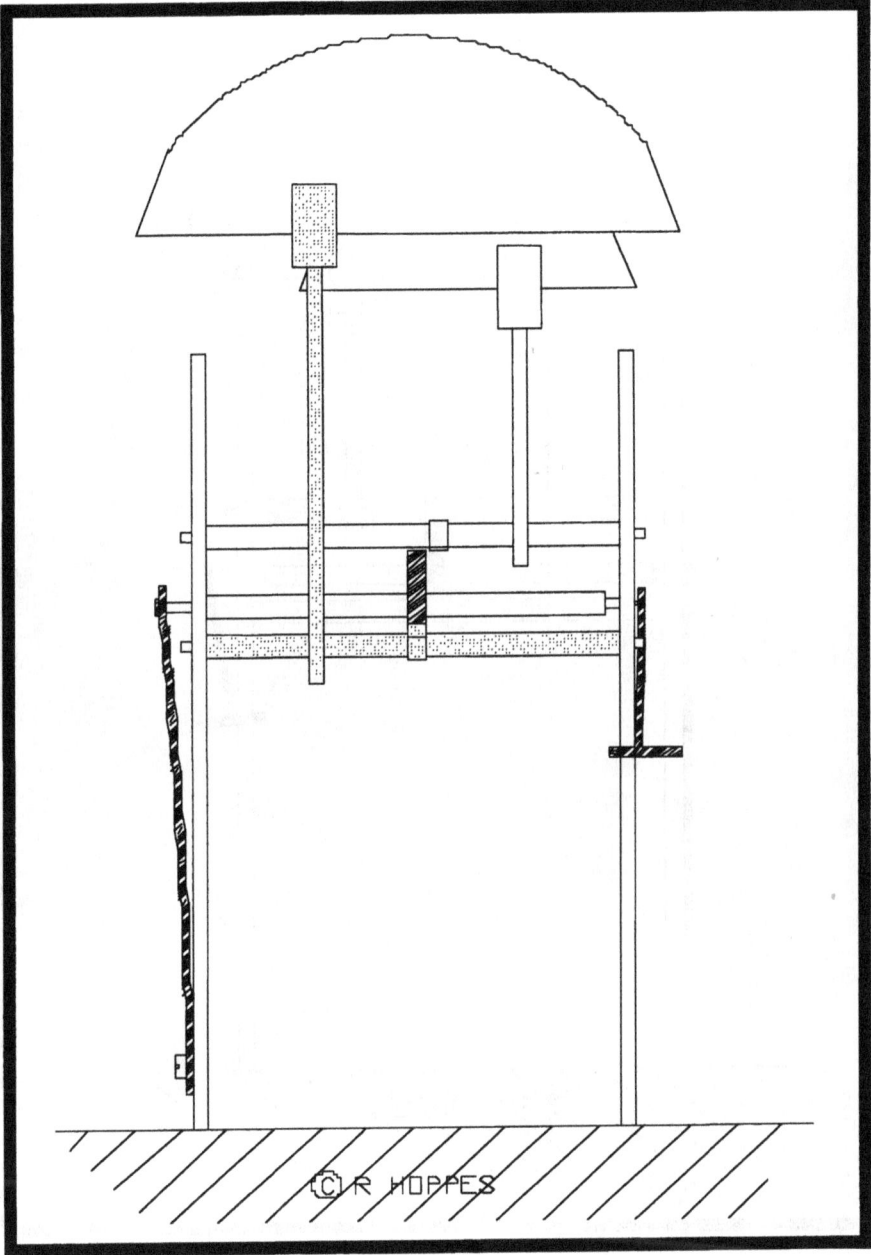

FIGURE 55. Shoe in Upper Position for Hour-Hammer Actuation

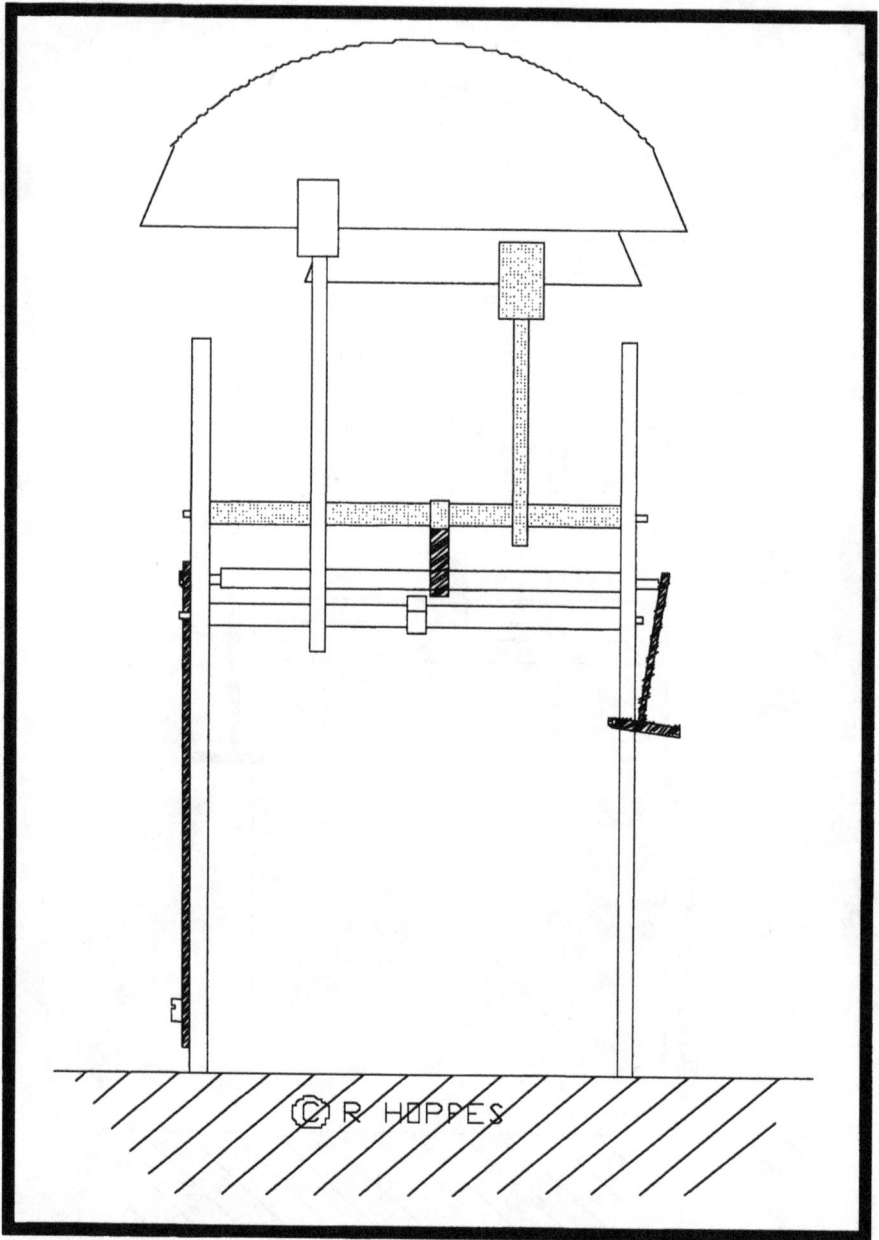

FIGURE 56. Shoe in Lower Position for Quarter-Hammer Actuation

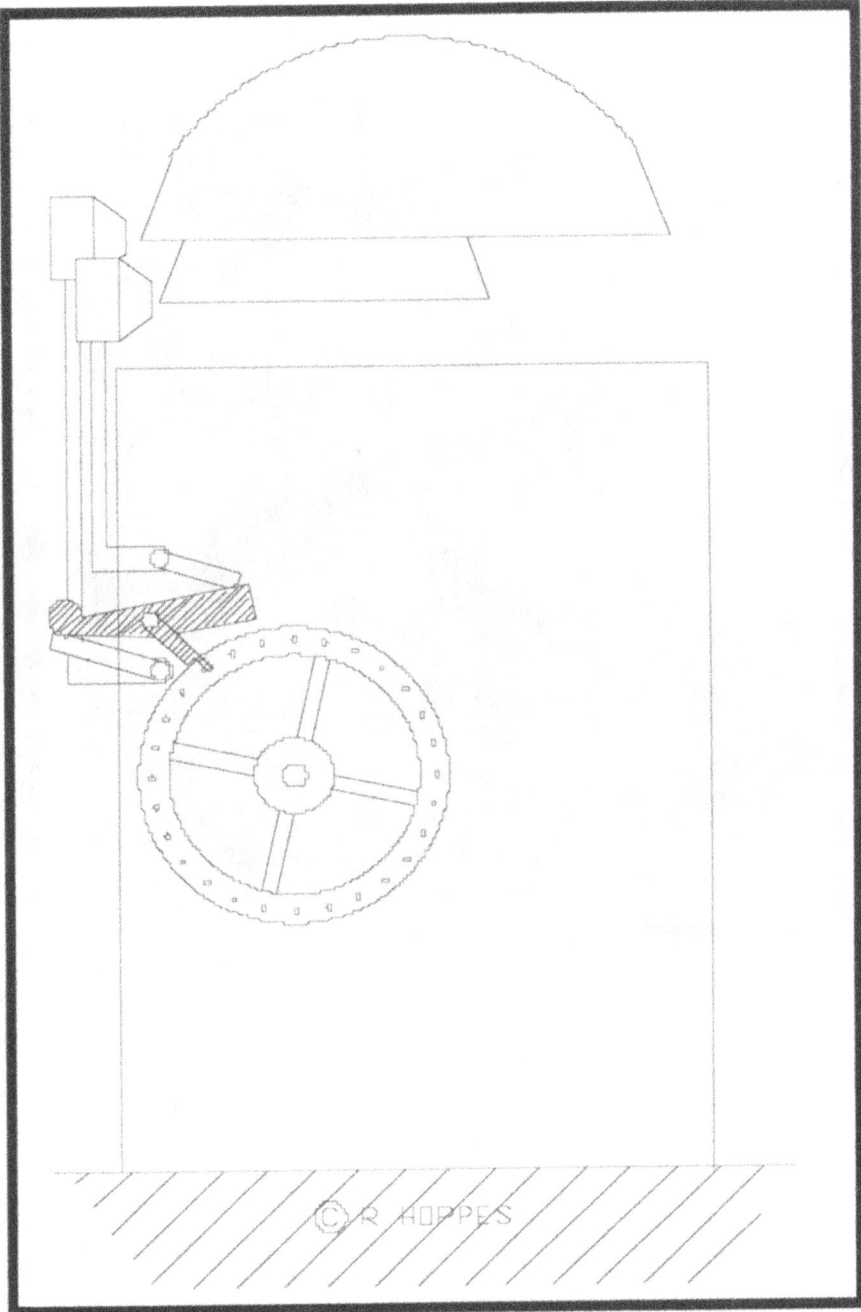

FIGURE 57.   Hammers, Intermediate Shaft, and Strike Wheel

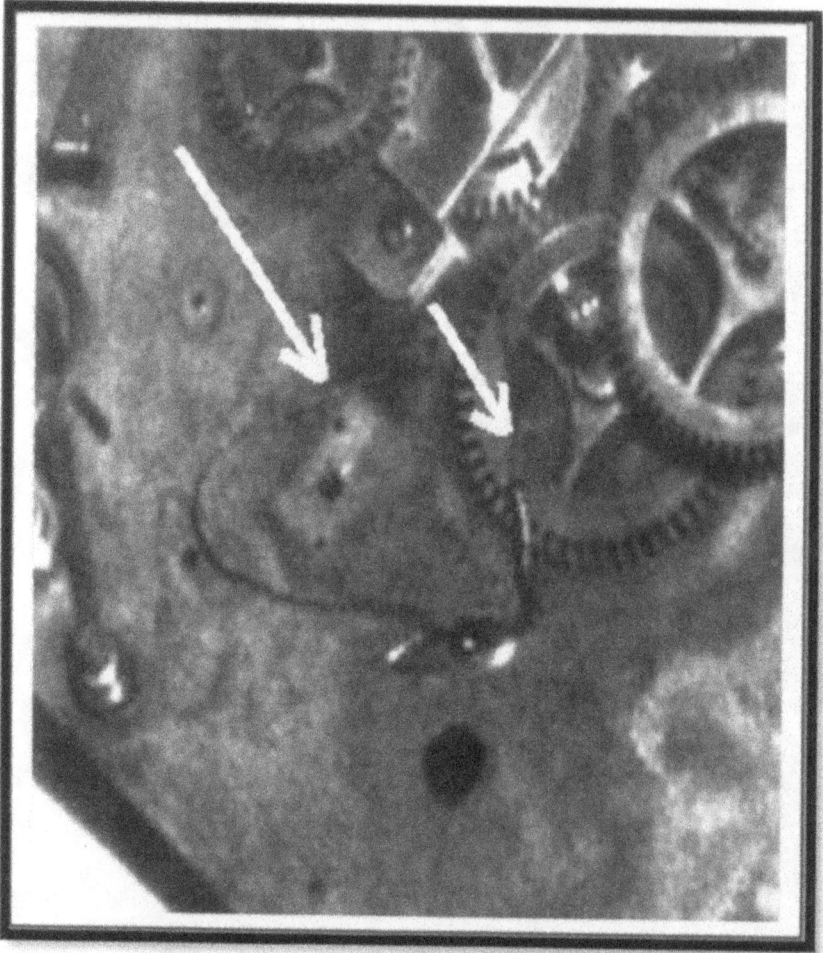

FIGURE 58.   Earlier Rack-Return Springs

7

The Music Train

L EFT AND RIGHT VIEWS OF the music drum, hammers, and bells are shown in Figure 59. The fly, which governs the speed of the tune play, is visible on the rear of the movement in the right-hand view.

FIGURE 59.   Left- and Right-Hand Views of the Music Drum and Bells

Figure 60 is a photograph of the gearing, tune trip, tune repeat, and tune change components. Figure 61 is a line drawing of the music train and controls.

The music section has a substantial number of settings to select from. These settings determine the characteristics of the tune-playing operation. They define how frequently a tune is to be played and the number of times a tune will be repeated before the performance is ended.

The music train can be set to be silent or engaged automatically at one of four selected intervals, Silent, 2 hrs, 1 hr, $\frac{1}{2}$, hr and $\frac{1}{4}$ hr. In addition, a tune

FIGURE 60. The Music Train Trip, Repeat, and Tune-Change Gearing

can be set to be repeated 1, 2, 3, or 4 times each time the play sequence is initiated. The weight drops 0.1374447" for each play of a tune. So the weight drop per day will increase as tunes are played more frequently or the number of repeats for each play is increased. The intervals between windings for a repeat selection of one, are listed.

The winding interval also becomes more frequent as the number of repeats is increased. Divide the winding intervals by the number of repeats selected per play. For two repeats divide the winding interval by 2, and by 3 for three repeats. Similarly for four repeats divide by 4, so if the 1/4-hour play interval is selected along with 4 repeats per play, the winding interval is basically every day (4.5 days/4 = 1.1 days).

Play every 2 hrs. (12 plays per day) = 36 days

Play every 1 hrs. (24 plays per day) = 18.1 days

Play every 1/2 hrs. (48 plays per day) = 9 days

Play every 1/4 hrs. (96 plays per day) = 4.5 days

Tunes may also be played at any time by the means of the Tune Repeat lever. Of course this also reduces the length of time between required windings.

The gear train of the clock movement also drives the tune-change snail and tune indicator. The gear train causes the 10-tune snail to make one revolution in 40 hours, causing a tune change every 4 hours. So, starting with tune 1 on day 1, the tune sequence is:

Day 1 the tunes are 1, 2, 3, 4.

Day 2 the tunes are 5, 6, 7, 8.

Day 3 the tunes are 9, 10, 1, 2.

Day 4 the tunes are 3, 4, 5, 6.

Day 5 the tunes are 7, 8, 9, 10.

Day 6 the tunes are 1, 2, 3, 4.

We see that the tune sequence repeats on day 6.

THE TRIP SEQUENCE

The trip mechanism is illustrated in Figure 61. The figure illustrates the music train gearing and control levers. The gears that change the tune selection are not shown so as not to clutter the diagram, and will be described later.

To better illustrate the operation of the play sequence, Figure 62 shows the first stage of the trip operation. It illustrates the pin on the trip wheel just as it makes contact with the trip pawl. Note that the toothed sector, which counts the number of tune repeats, is still in its most left-hand position. It is held in this position by the pin of the warning lever. This pin on the warning lever is shown as a solid dot in the figure. When the warning lever is lifted by the trip pawl, the sector falls to the right, influenced by the spring pushing against its curved tail below its pivot point.

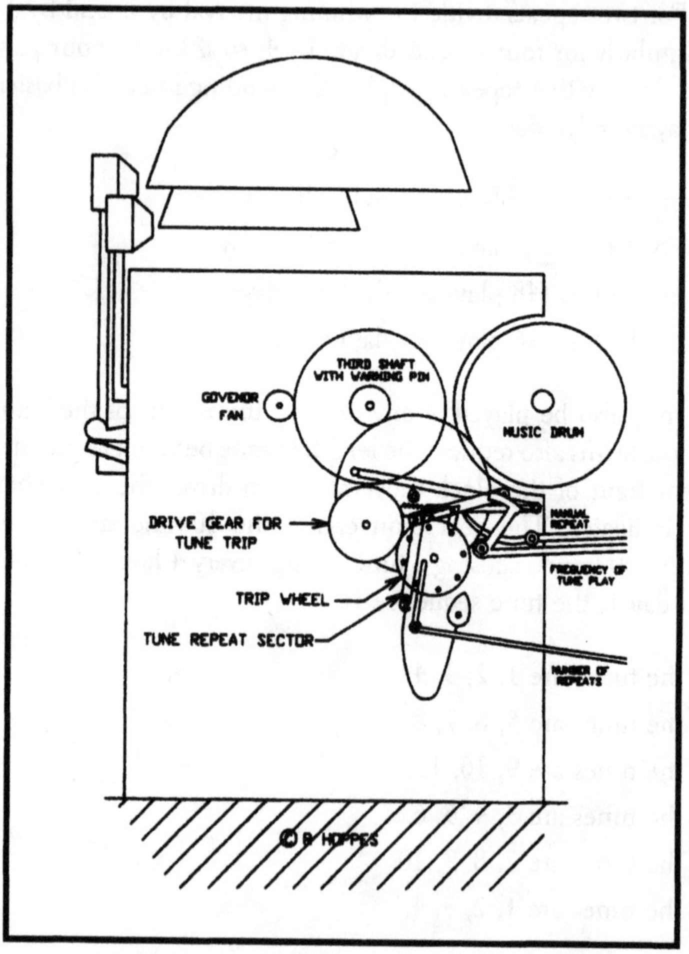

FIGURE 61.   The Music Train and Controls

Figure 63 illustrates tip of the trip pawl ready to escape from the lifting action of the trip pawl. Note that the trip pawl has lifted the trip lever, which in turn has lifted the warning lever. Also note that the trip pawl has rotated slightly clockwise on its pivot screw during the lifting action of the trip pin. As soon as the trip pawl escapes, the leaf spring mounted above it causes the pawl to return to its previous position, which results in a small counterclockwise rotation. This action allows the tip of the trip pawl to pass over top of the pin and clear it, so that the trip pawl is free to fall back to its starting position.

This action releases the warning lever, allowing it to drop clear of the pin on the third shaft gear. The gear train is now free to run allowing the tune to play.

Note that when the warning lever was lifted, the toothed sector moved to the right, coming to rest against the tip of the lever that selects the number

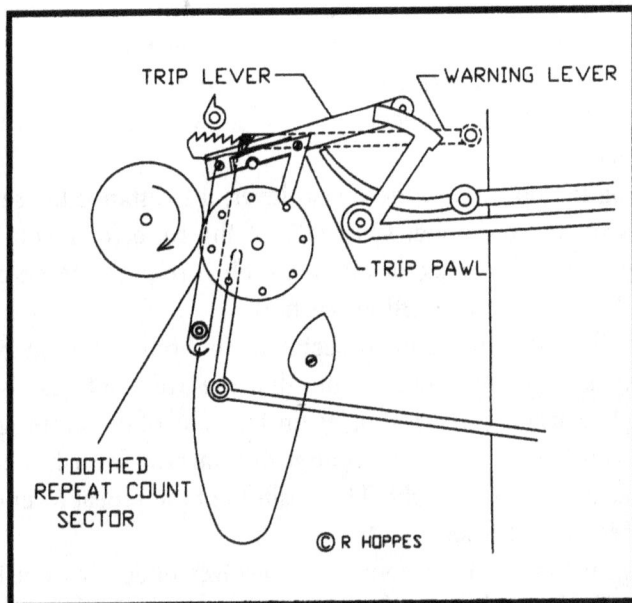

FIGURE 62. Start of Tune Trip Sequence

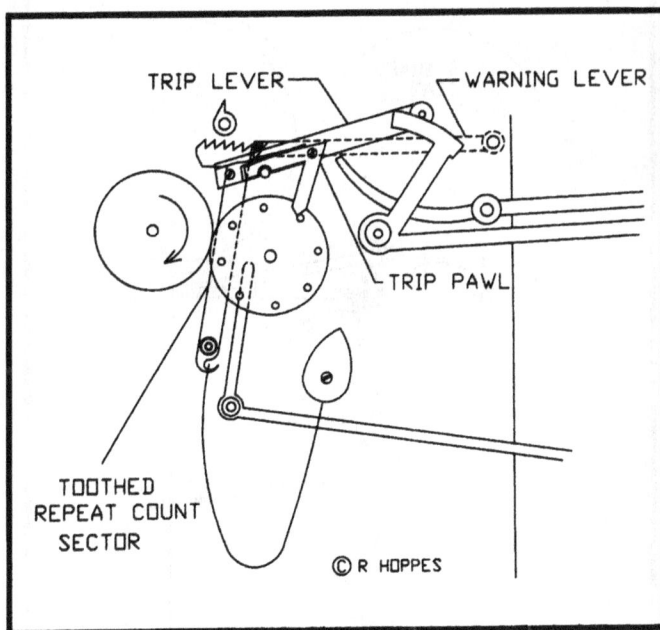

FIGURE 63. Tune Trip in Warning Phase

of tune repeats. The lever tip contacting the sector arm is illustrated in phantom because it is located under the trip wheel shown in Figure 63.

The repeat selection is better illustrated in the following two illustrations, Figures 64 and 65. The tune-selector lever is shown in black, and the toothed sector is shaded.

By changing the position of the repeat lever, the distance the sector moves to the right can be controlled. One play of the tune (or revolution of the drum) is required for each sector tooth, which is required to be gathered so as to return the sector to its starting position.

Figure 64 illustrates the toothed sector in its home position prior to the warning lever having been raised. Note that the 3rd shaft gear is prevented from rotating because its pin is resting against the end of the warning-lever arm.

Figure 65 illustrates the condition after the warning lever has been raised. The sector has moved to the right. The position of the repeat lever determines how far the sector is allowed to fall.

A gathering pallet on the second shaft revolves once for each turn of the music drum. Each revolution advances the sector to the left by one tooth. The pin of the warning arm acts as a ratchet pawl, preventing the sector from falling back against the repeat-selector lever.

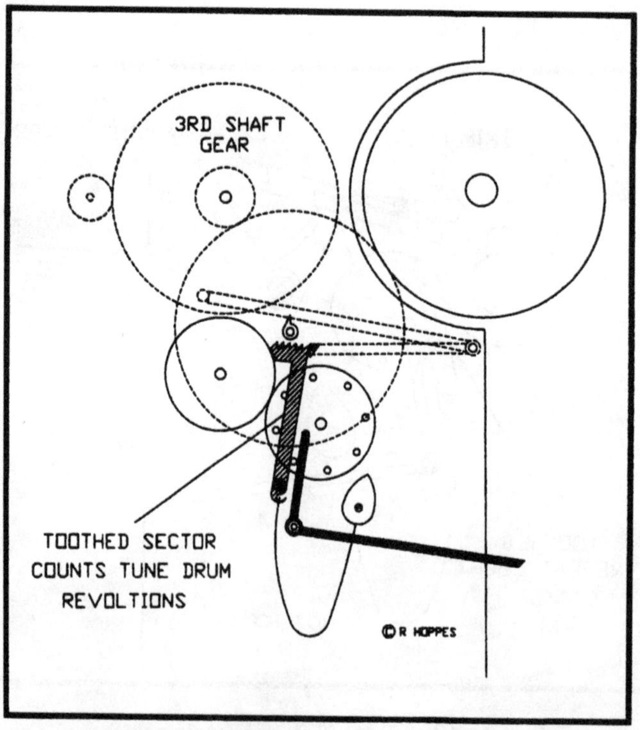

FIGURE 64.　Tune Repeat Selector in Its Home Position

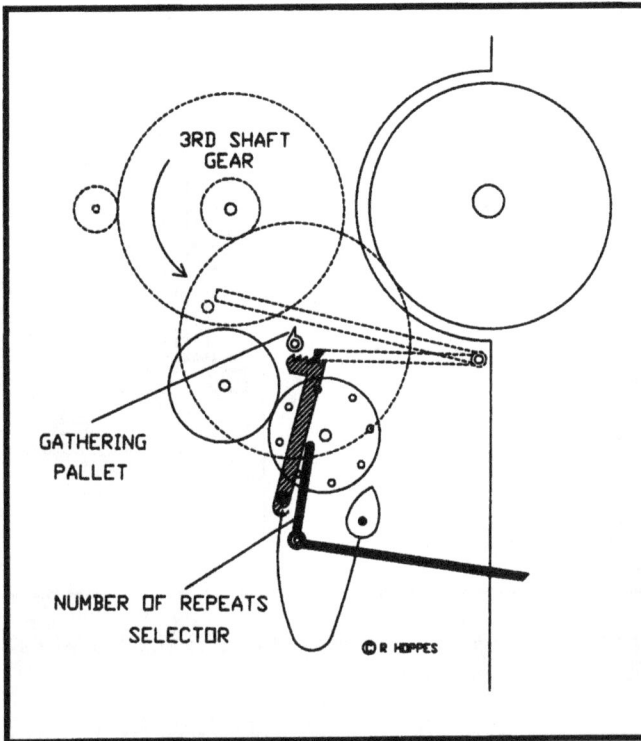

FIGURE 65. Tune Repeat Selector in the Two-Repeat Position

The tune will no longer be repeated after the pin of the warning lever passes the last tooth of the sector. After the last tooth has passed, the sector teeth no longer support the warning lever, so it drops to a lower position. This lower position does not allows the pin on the third shaft gear to pass, ending the tune-play sequence.

How often music plays is determined by the position of the tune-play selector. The manner in which the selection is accomplished is illustrated in Figures 66 and 67. The location of the different-length pins on the wheel, in conjunction with the position of the trip lever, determines the interval of the tripping action.

The trip wheel revolves once in two hours, and has pins of different lengths. By changing the distance of the trip pawl from the trip wheel, the frequency at which the trips occur can be changed.

The longest pin causes a trip once per revolution of the wheel or every two hours when the trip pawl is in the 2-hour position. One slightly shorter pin on the opposite side of the wheel in conjunction with the long pin will trip the tune play every hour when the pawl is in the 1-hour position. The next shorter pins consist of two halfway between the previous pins, allowing a trip

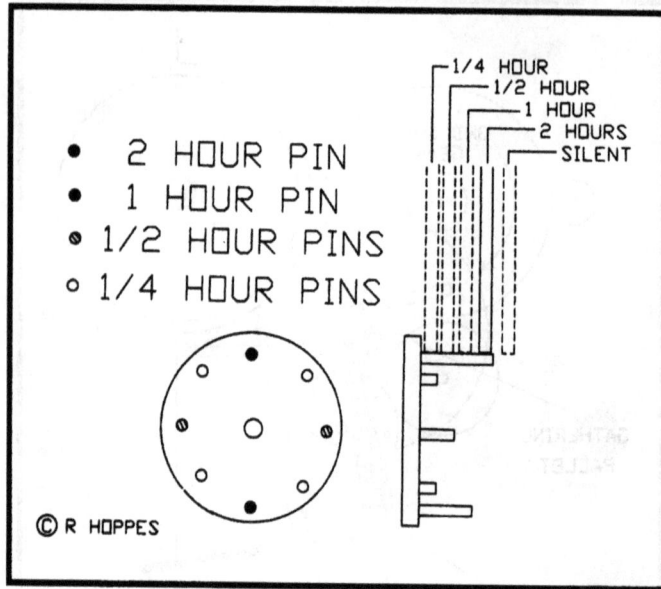

FIGURE 66.   The Trip Pins and the Five Tip-Lever Positions

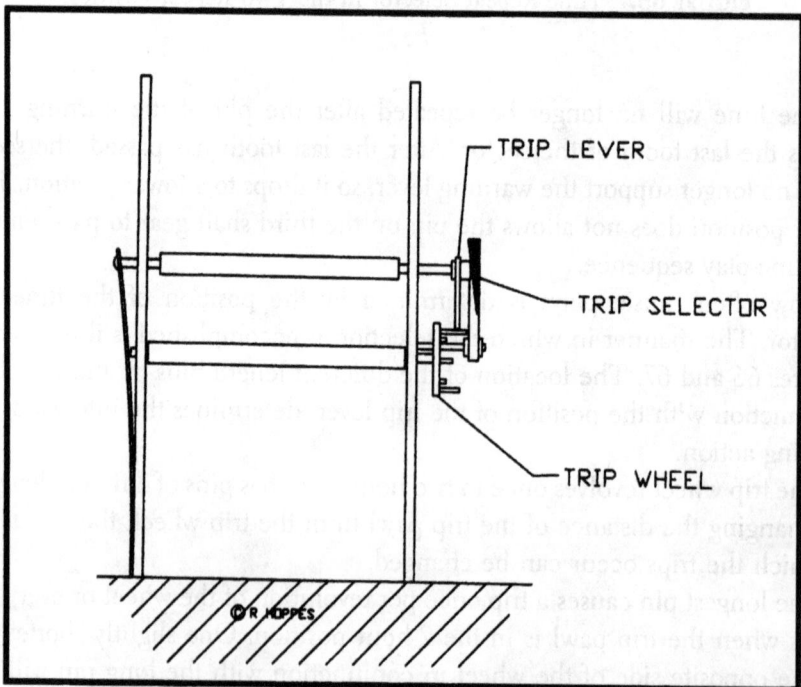

FIGURE 67.   Method of Tune Trip Selection

every $1/2$ hour. There are also four additional short pins between these, for $1/4$-hour trips.

TUNE-CHANGE GEARING

The gearing that changes the tunes is illustrated in Figure 68.

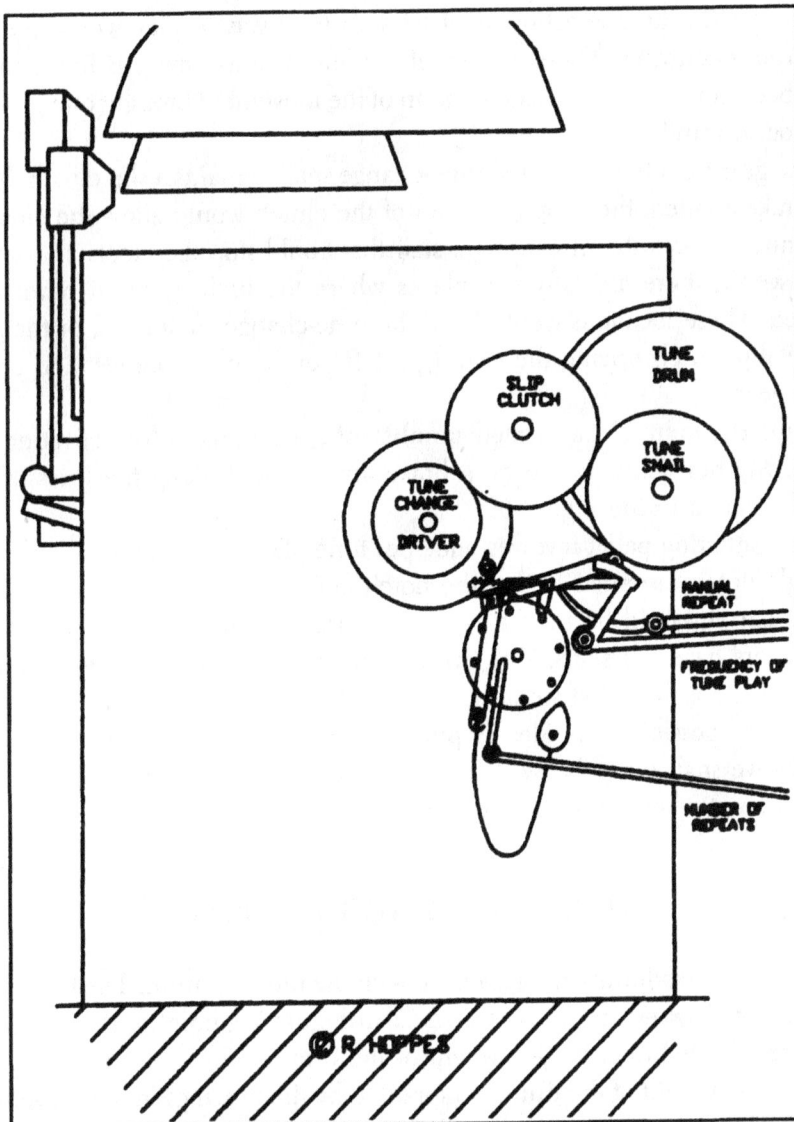

FIGURE 68.   The Tune-Change Gear Train

Because the clock movement is continually advancing the tune-change snail, a barrel shift to cause a tune change can occur at any time, even during a tune play.

The dangers of shifting a music barrel during play are well known and understood by all individuals familiar with mechanical music pinned on a drum. Allowing this to occur will eventually cause bending and possibly breakage of pins on the barrel, because pins will eventually collide sideways with a hammer-lift tail. This type of collision will cause pins to be bend or break.

It is possible that this problem is the result of someone removing an important part of the music train. A brake mechanism that would impede changing the position of the tune-selection snail while a tune was in play would prevent this from occurring. There are few if any clues as to how this brake might have been constructed. An examination of the movement brought one possible method to mind.

The gear train that drives the tune-change snail contains a slip clutch. With the brake applied, the slipping action of the clutch would allow the clock to continue to run without causing a stall that could stop the clock.

However, there are only two places where the braking could have been applied. These locations would be at the tune-change snail itself, or the gear of the slip clutch driving the snail (gears R4 or R3 in my description of the clock gearing).

Similarly, there are a limited number of good choices for activating the brake. My best guess would be a brake arrangement along the lines of the one illustrated in the Figure 69.

The gathering pallet revolves once per tune play. So mounting a disk with a single notch, oriented so that the notch corresponds to the home position of the barrel, could be used to activate a brake. Both slip-clutch and tune-snail gears have flat skirts. The cylindrical surfaces of these skirts would be ideal for use as breaking surfaces.

Though possible, this idea is pure speculation, and whether it has any validity we shall never know. It is the simplest solution I can envision. The correct answer appears to be lost to history.

MUSIC DRUM ADJUSTMENTS

There are three adjustments associated with the musical drum. The two down-ward-pointed arrows, numbered 1 and 2 in Figure 70, point to thumb screws that are adjusted to carry the weight of the drum. When adjusted correctly, none of the weight of the drum is supported by the bearings in the movement plates. The weight of the drum will be balanced against the elasticity of the supporting suspension by the thumb screws.

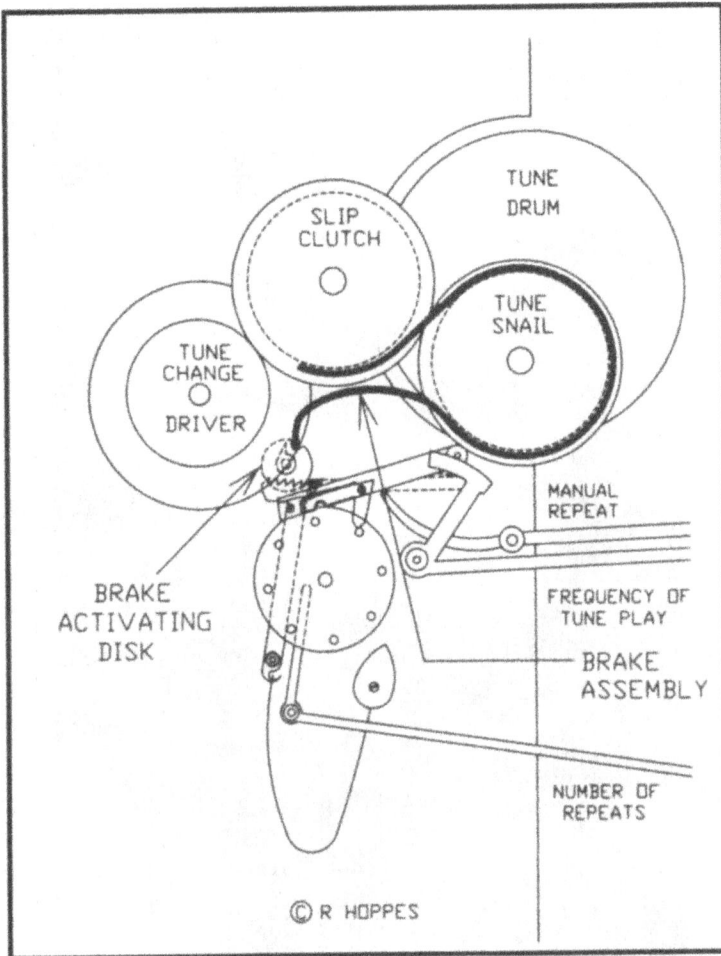

FIGURE 69. A Possible Braking Arrangement

With the weight of the drum on the long supporting wires, the drum can swing freely front to back within the movement frame. This freedom minimizes the sliding bearing friction during tune changing. It is not intended to adjust the amount of hammer lift required for tune play. Hammer lift is established by the alignment of the bar carrying the hammers with the periphery of the drum.

The third arrow, number 3, pointing upward, adjusts the force of the hammer-return springs. It is effectively a volume control for the striking of the bells. Greater return-spring force requires a greater driving force to the barrel. Too large of an adjustment may stall the rotation of the drum, as the rotational force available is limited by the mass of the weight, driving the music train.

FIGURE 70. The Music Drum Adjustments

8

Setting the Dial Indications

PREPARATION

To prepare to set the dial indications, the first task is to determine the angular position for either the date, or some other reasonably recent date, on which the indications are to be set. The angular positions in the zodiac for each of the indications need to be determined and entered in the setting list in Table 12. Figure 71 illustrates the angular ecliptic heliocentric positions in the ecliptic to which the astrological positions are to be referenced.

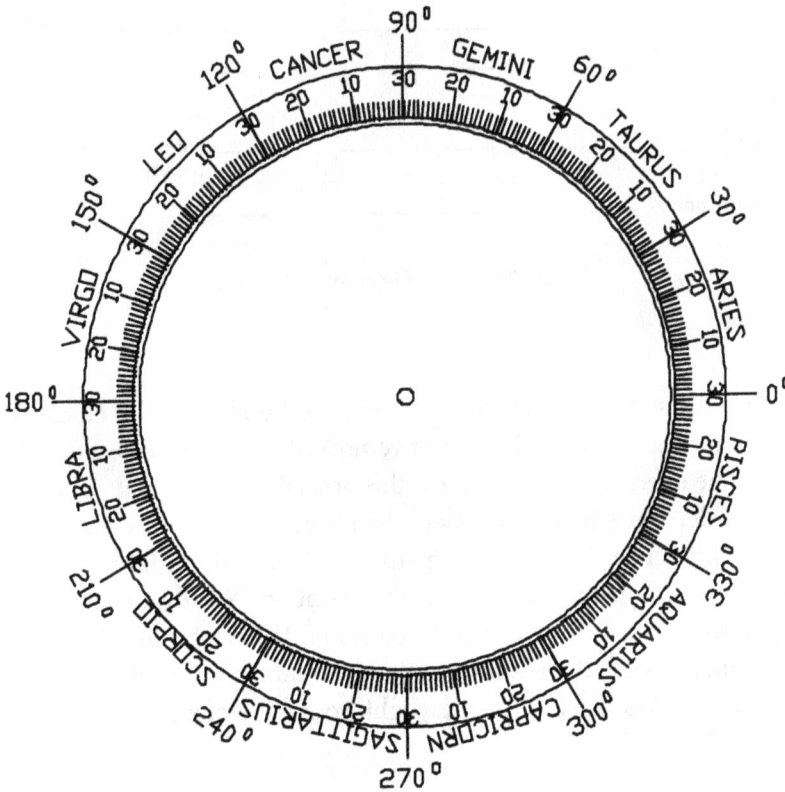

FIGURE 71. Angular Ecliptic Heliocentric Positions in the Ecliptic

Table 12. The Drexel Rittenhouse Clock Dial-Setting List

Reference date and time for settings:

Year_____ Month_____ Day_____

Time: Hr_____ Min_____ AM_____ PM_____.

Position of	Angle	Zodiac Sign	Degrees into Sign
Saturn	_____	_____	_____
Jupiter	_____	_____	_____
Mars	_____	_____	_____
Earth	_____	_____	_____
Venus	_____	_____	_____
Mercury	_____	_____	_____
Position of			
Sun	_____	_____	_____ (Sun = Earth ± 180°)
Moon	_____	_____	_____
Moon's Orbit			
Perigee	_____	_____	_____ (Point Nearest the Earth)
Ascending node	_____	_____	_____

	Phase (Quarter)	Days into Quarter
Moon phase	_____	_____

Now begin the setting session by removing the hood from the clock to gain access to the rear of the dial. Do not remove the dial because the calendar hand must be operational to complete this procedure. Note that no indication on the dial denotes whether the clock hands are indicating AM or PM, and there is a difference. When passing from PM to AM, 12:00 midnight, the paddle on the D1 gear advances the date indication. No advance occurs when passing from AM to PM, which is 12:00 noon. When starting the clock and placing it into service, the advance of the date indication needs to be verified. If the advance takes place during daylight hours, the clock hands need to be advanced 12 hours.

THE CALENDAR HAND

Adjust the calendar indication so it corresponds to the date represented by the planetary data. The calendar hand is set to the corresponding date by use of the orrery crank on the shaft at the lower right of the sun–moon dial shown in Figure 3.

THE EQUATION-OF-TIME DIAL

Note that the required angular gear and arm positions of the equation-of-time mechanism can only be set with the dial removed from the movement. If the positions are incorrect, it is necessary to disassemble the mechanism and reassemble it to the requirements illustrated in Figure 35.

The equation-of-time dial must now be set to correspond to the date of the calendar hand. Because some positions of the hand are repeated several times during the year, the proper relationship of the mechanism for the indicated date must be verified. This is best done by selecting a value that is indicated only once per year. There are two such indications, +14.4 (Sun Fast 14.4 min on Feb 11) and −16.4 (Sun Slow 16.4 min on Nov 3). By means of the orrery crank, place the calendar hand on one of these dates. If the equation-of-time dial does not indicate the correct value, disengage the H3 gear (Figure 20) and rotate the orrery gearing until the equation-of-time reading is correct. This establishes the proper phase relationship of the calendar hand and the equation of time. Re-engage the H3 gear and now move the calendar hand to May 19 (−3.7 min) or Dec 30 (+2.5 min) and verify the indication is correct. Again disengage gear H3 and make a second calibration correction if required. The equation-of-time values for each day of the year are shown in Table 11.

THE ORRERY PLANETS

Set the calendar hand, using the orrery crank, to the date that corresponds to the recorded planetary positions of Table 12, the Clock Dial-Setting List.

The arm that carries the planet Saturn is held in its angular position by a pin in the end of the tube of the P1 planet gear. To set the position of the planet arm, remove the screw on the upper end of the bridge for the gear cluster P0 through P6. This will allow the cluster gear to be disengaged from

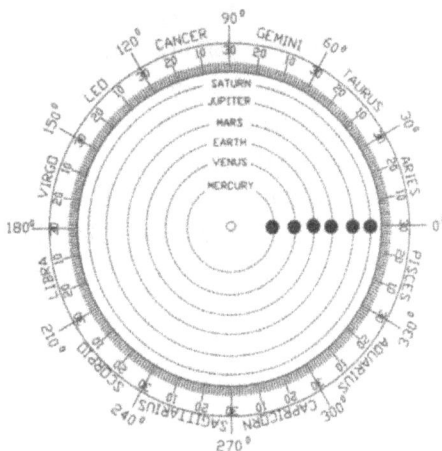

FIGURE 72.   The Orrery Dial

the S1 (Saturn) gear. The position of the Saturn arm can now be set and the gear reengaged. Be careful not to alter the engagement position of the E1 gear, which would change the equation-of-time setting, now replace the bridge screw.

The remainder of the planet arms are friction fitted, similar to the hour hand of many clocks. Unscrew the center knob representing the Sun and then remove the planet arms one at a time. Replace them in reverse order (Jupiter, Mars, Earth, Venus, Mercury, and then the Sun) in their proper zodiac positions. Attempting to set the planet positions by disengaging and reengaging the gearing will allow only whole-tooth positions to be selected. The tooth positions will not always allow the planets to be positioned with sufficient accuracy. Over time, the rate of the orrery planets tends to be slow rather than fast. So any positioning errors should favor an early indication rather than a late one. The planets rotate in the counterclockwise direction. The orrery dial with its planets at the reference position of 0° is illustrated in Figure 72.

THE SUN–MOON DIAL

Disengage gear S2 (by removing the tapered pin securing it) and adjust the position of the sun. Similarly, disengage gear L12 and adjust the position of the moon indicator. These indications also tend to loose rather than gain, and rotate in the counterclockwise direction. Replace the gears and taper pins. Now adjust the rotational phase of the moon on its indicator by removing the arm with the moon from its shaft and positioning the moon to the proper phase indication. This arm is attached via a tapered pin on the dial side of the shaft. The direction of the moon's rotation moves in the same way as one would turn a right-handed screw to tighten it. The sun-moon dial is illustrated in Figure 73.

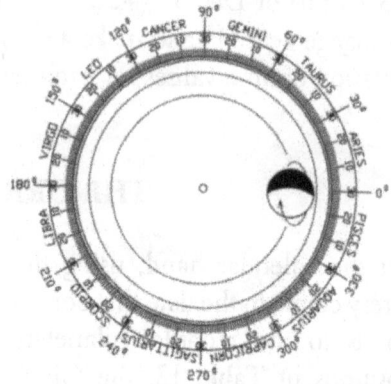

FIGURE 73. The Sun–Moon Dial

THE MOON ECLIPTIC NODE INDICATOR DIAL

The apogee (farthest from the Earth), which
is the end opposite the perigee (perigee
± 1800), is easier to set. Disengage gear L7
(by removing the taper pin on its shaft) and
adjust the positions of the ecliptic indicator
by rotating L10 (it is the one closest to the
dial). Also adjust the node indicator by rotating L9 (the one farthest from the rear of
the dial). Reengage gear L7 and replace its
taper pin. The moon ecliptic node indicator
is illustrated in Figure 74.

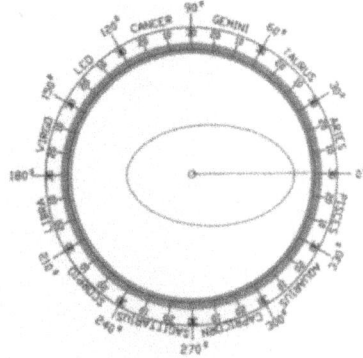

FIGURE 74.   The Moon Orbit Dial

THE LUNAR DIAL

Rotate the friction-driven
moon-phase indication to the
proper position by applying
hand pressure to the moon
disk. The markings are for the
number of days before and
after the occurrence of a new
and full moon. The dial may
be rotated in either direction.
Note that the motion of the
dial is clockwise. The lunar
dial is illustrated in Figure 75.

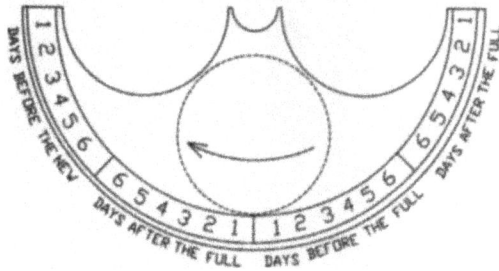

FIGURE 75.   The Lunar Dial

RESTARTING THE CLOCK

Using the orrery crank, move the calendar hand so it corresponds to the current
date. If the date in the aperture is not correct, reach behind the dial and rotate
the indicator ring by hand to the proper date. Replace the hood, set the time,
and restart the clock.

NOTES

1. The moon requires that it be set exactly because it moves $1/2$ degree every 1 hour.
2. The two-moon disk in the center of the dial rotates counterclockwise.
3. See Figure 9 on page 13 for the daily equation-of-time correction.

Index

9 781606 189924